ROUTLEDGE LIBRARY EDITIONS:
ISLAM, STATE AND SOCIETY

I0130590

Volume 6

SCIENCE, TECHNOLOGY AND DEVELOPMENT IN THE MUSLIM WORLD

SCIENCE, TECHNOLOGY AND DEVELOPMENT IN THE MUSLIM WORLD

ZIAUDDIN SARDAR

Routledge
Taylor & Francis Group

LONDON AND NEW YORK

First published in 1977 by Croom Helm Ltd

This edition first published in 2017
by Routledge
2 Park Square, Milton Park, Abingdon, Oxon OX14 4RN

and by Routledge
711 Third Avenue, New York, NY 10017

Routledge is an imprint of the Taylor & Francis Group, an informa business

British Library Cataloguing in Publication Data
A catalogue record for this book is available from the British Library

ISBN: 978-1-138-23270-9 (Set)
ISBN: 978-1-315-31161-6 (Set) (ebk)
ISBN: 978-1-138-21976-2 (Volume 6) (hbk)
ISBN: 978-1-138-21978-6 (Volume 6) (pbk)
ISBN: 978-1-315-41453-9 (Volume 6) (ebk)

Publisher's Note
The publisher has gone to great lengths to ensure the quality of this reprint but points out that some imperfections in the original copies may be apparent.

Disclaimer
The publisher has made every effort to trace copyright holders and would welcome correspondence from those they have been unable to trace.

SCIENCE, TECHNOLOGY AND DEVELOPMENT IN THE MUSLIM WORLD

ZIAUDDIN SARDAR

CROOM HELM LONDON

© 1977 Ziauddin Sardar
Croom Helm Ltd, 2-10 St John's Road, London SW11

British Library Cataloguing in Publication Data

Sardar, Ziauddin
    Science, technology and development in the Muslim
    world.
    1. Technology – Social aspects – Islamic countries
    I. Title   II. Muslim Institute
    301.24'3'0917671   T14.5

    ISBN 0–85664–554–0

to the *Ummah*

Reproduced from copy supplied
printed and bound in Great Britain
by Billing and Sons Limited
Guildford, London, Oxford, Worcester

CONTENTS

PREFACE

My aim in writing this book has been simple: to present a Muslim view of development and to highlight some of the related issues now being debated in the Muslim world. There is no lack of work on development representing various normative positions including the conventional wisdom of both left and right. To the best of my knowledge, this is the first book which specifically sets out to present a Muslim point of view.

How we perceive development depends, to a large extent, on our world-view. To establish the proper context, therefore, it will be necessary to outline the parameters of the Muslim world as well as the Muslim world-view. The nature of the subject demands a somewhat theoretical treatment of science, science policy and Muslim culture. Of course, this treatment derives from my own normative position, as does the analysis presented in the later chapters: one is almost forced into this for the literature on development derives largely from a narrow economic and materialistic view and overlooks the cultural and ethnical dimensions of development.

I have tried to write a general book not focusing on any particular Muslim country. As such I have drawn my examples from various appropriate Muslim countries and, indeed, I have also cited examples from non-Muslim countries where their experience and conditions are pertinent. The over-view treatment of development in the Muslim world has been, to some extent, at the cost of depth. I am fully aware of this limitation. Nevertheless, I feel that at this point it is more important to establish a general view of science, technology and development in the Muslim world than to concentrate on specific issues and particular countries.

Many people have helped directly and indirectly in the preparation of this book. I would like to thank specifically my friend Dawud G. Rosser-Owen for his invaluable help with Chapter 3. Thanks are also due to my brother Jamal Sardar for his comments, often biting, on various drafts of this book. And I am particularly grateful to my friend Dr J.R. Ravetz who read the entire manuscript and made some valuable comments and suggestions for the improvement of the text. Finally, thanks are also due to M.H. Faruqi for all his moral support.

Any faults and discrepancies in this book are, of course, all mine;

and I bear responsibility for them. However, if the book, with all its limitations, is of value to the many Muslims concerned about present development strategies, and indicates a general direction for viable Muslim alternatives, I shall have achieved my purpose.

Ziauddin Sardar
9 September 1976
London

INTRODUCTION

The apparent conflict between science and religion, which originated in the Middle Ages of Christendom, is now increasingly outmoded. Today, the drift is towards synthesis rather than conflict. This book is an attempt to show how Muslims can assist this synthesis in their contribution to the development of science and technology and their own society.

To comprehend the ideas and arguments in this book it is necessary to have some understanding of Islam and and thus an appreciation of the Muslim point of view. A brief and lucid exposition of Islam, in the words of Muhammad Hashir Faruqui is helpful:

> In simple and precise terms Islam is the belief in the oneness and supremacy of Allah and a willing obedience to His commands. Man has been created in a state of natural goodness, given the faculty to discriminate between right and wrong, and the freedom to choose either. Not left alone to grope in instinct and desire, God provided man with guidance through His authorised representatives — prophets and messengers, to teach and to demonstrate through personal example and practical guidance. These individuals were given the faculty and the felicity to invite people to that which is good but not to coerce or to impose.
>
> Since God is not a part-time deity nor one who shares his powers with one or more of his kind, the guidance He provides is comprehensive, universal and eternal. He would be a very selfish God who knew everthing but told his best creation, Man, nothing but how to praise Him and not help him in the conduct of his life, in the home, in the market, in the courts and in the parliament. If you believe in an Almighty, all-Knowing, Wise, Kind, God you must logically accept Him as *the* source of guidance. And· since God is God in his own right and not because we appoint or elect him to be so, therefore, His knowledge and authority and guidance are not time or situation bound, these are absolute and eternal. They are absolute and eternal, but one is free to accept or not to accept although after having recognised Him as *the* Deity, one is downright silly to say that, look

9

God, you are not progressive enough, what you told Muhammad 1,400 years ago is now old and out of date.

The position of man, according to Islam is that of a *de jure* vicegerent and a *de facto* sovereign, i.e. he is required to discharge his commission as a caliph of God but he actually has the freedom to behave as a sovereign, though that would be without lawful authority.

This freedom and autonomy given to man is not unlimited. It has a life-time tenure, after the expiry of which one is accountable for the use or misuse of this lease of independence. This freedom, this tenure, and then the accountability, the judgement, the reward, the punishment and the hereafter, all these are an integral, indeed a logical part of the scheme of things. This is the scheme of Islam. One may accept it or deny it. One may like it or not, the scheme is there, it operates, nevertheless.[1]

This then, is a view of Islam, the essence of Muslim behaviour. It is difficult to understand the ethos, the aspirations and the hopes of Muslims without appreciating Islam. It is just as difficult to understand Islam by observing the behaviour of individual Muslims and Muslim societies. Often the theory and the practice do not tally.

This was particularly so during the first half of this century when the Muslim world was, by and large, a colonial area controlled by various Occidental powers. By subjection, occupation, mandated responsibility, special privileges, and spheres of influence the Occident largely directed the processes of political, economic and technical development.

A principal outcome of this authority was that scientific and technological development became very largely an Occidental concern. This opportunity was profitably exploited by predominantly Occidental entrepreneurs in such commodities as Iranian and Arab oil, Malayan and Indonesian rubber, tin, and a host of other primary resources. In this exploitation perhaps the greatest profit and misery were derived from the wholesale military conquest in the 1920s and 1930s of the independent Muslim states of Central Asia by the Communist Russians and Chinese. These, moreover, took their imperialism a stage further than anyone else and annexed their conquests.[2]

After the Second World War, the situation in the Muslim world began to change. Firstly, in 1947 the independent Muslim nation-state of Pakistan emerged from the former British Indian Empire. This was a watershed for the colonized Muslim communities.[3] Before 1939, the only independent Muslim countries were those in the Levant which were either part of the Occidental sphere of influence (like Egypt, Iran, Iraq, Syria, the Lebanon, and Transjordan) or else had been severely weakened in the First World War (Turkey and Saudi Arabia). These two countries were primarily concerned with the problems of nation building and reconstruction and were not able to play a major part in the international affairs of the Muslim world. The creation of Pakistan provided a new dynamic within the Muslim community so that now there are forty-three independent nation states. However, after the Second World War a strange mutant type of society emerged that was partially Muslim in tradition and values but Occidental in behaviour, modes of thought and even outlook. The colonial domination had altered the character of many Muslim societies.

Naturally, in their initial development, the Muslim countries looked up to their former colonizers. The new states followed the principles of 'economic development', 'advancement', 'progress', 'industrialization', 'democracy' — in short, the Occidental model of a modern state.

In this development, most Muslim nationals were only permitted a minor role and were only trained for the more modest positions. The chief concern of the new elite left in power by the departing masters was to share in the profits of development and to control it. There was little ambition to manage new enterprise and much less to initiate it.

Now that the Muslim world has matured both politically and technologically, this state of affairs no longer holds. The Muslim world is now not content to leave its destiny in Occidental hands; it wants to produce its own oil, manufacture its own cloth, and operate its own utilities. Throughout the Muslim areas there is a new determination towards self-development and self-help.[4]

Indeed, scientific and technical development is now considered, in some Muslim countries, to be a concomitant of political freedom. It has become a national aspiration, an indispensable condition for sovereignty and dignity. Muslims are beginning to see that to be free yet dependent on aid, to be free yet dependent on Occidental technical skill, to be free yet dominated by Occidentalized elites, is not real freedom. Muslims are no longer prepared to accept a one-way transfer of ideas from the Occident to the Muslim world, nor will they tolerate the old style of cultural bond.

Interaction between the Muslim world and the Occident must now be based on mutual respect and a dialogue between equals. Both must seize the opportunity to strive as never before to understand each other. We hope this book makes a small contribution towards promoting that understanding.

NOTES

1. M. H. Faruqui, 'Revolution or Renovation?', *Impact International Fortnightly* 2, 14 8-9: for a detailed exposition see A. R. Azam, *The Eternal Message of Mohammad* (Devin Adair, New York, 1964); M. Hamidullah, *Introduction to Islam* (Centre Culturelle Islamique, Paris, 1959); M. Qutb, *Islam: The Misunderstood Religion* (Darul Bayan, Kuwait, 1967); and Khurshid Ahmad (ed.), *Islam — Its Meaning and Message* (Islamic Council of Europe, London, 1976).
2. See Sir Olaf Caroe, *Soviet Empire* (Macmillan, London, 1967); and Haider Bammate, *Visages de l'Islam* (Paris, 1955).
3. See Kalim Siddiqui, *Conflict, Crisis and War in Pakistan* (Macmillan, London, 1974) and *Functions of International Conflict; A Socio-Economics Study of Pakistan* (Royal Books, Karachi, 1975); Khalid bin Sayeed, *Pakistan: The Formative Phase 1857-1948* (Oxford University Press, London, 1968); and K. K. Aziz, *The Making of Pakistan* (Chatto and Windus, London, 1967).
4. 'Muslims the world over are impatient with their present conditions; they are in a ponderous and defiant mood. They want to stop and reverse the process of continuous economic, social, political, cultural and intellectual stagnation and humiliation that has been theirs for at least three hundred years. They are disillusioned with the traditional leadership which has reduced Islam to the level of any other religion concerned merely with personal piety and the Hereafter; they are equally disillusioned with the 'nationalist' leadership of the colonial era. They are in search of a new sense of direction; in short, a new destiny.' (Kalim Siddiqui, *Towards a New Destiny*, Open Press, London, 1974, p. 5.)

1 WHAT FORMS THE MUSLIM WORLD?

The world is seen by many to be divided into two groups: those who have 'too much' and those who have 'too little'; one rich, one poor; one overfed and overweight, one hungry and undernourished; one affluent and consumption orientated, one poverty-stricken and survival orientated; one 'developed', one 'developing'.

One world is then subdivided into two great hostile *blocs*: the supposed Free World (North America, Western Europe, Japan and Australasia) who in being free must therefore embody all that is just, democratic, and unrestricted, and the Communist Bloc of the Soviet Union and Eastern Europe, which is its antithesis. Depending on one's perspective, these blocs are either oppressive, totalitarian, and severely restrictive, or capitalistic, reactionary and imperialistic.

In addition, in recent years, the new concept of the *Third World* has emerged.[1] Frantz Fanon, who coined the phrase, perceived that the 'too little' countries in fact constituted a third power bloc. Although the two industrialized blocs might be in a position of mutual conflict, the third bloc of countries was so far removed from them in economic terms as to constitute a veritable Third World. From a purely materialistic point of view this is so, but examined more closely there is little similarity between Catholic Peru and Bolivia, Muslim Morocco and Somalia, Buddhist Thailand and Burma, Hindu India, Zionist Israel, and Communist Vietnam, except perhaps in a common economic non-development.

The American journalist William Buckley divided the world into 'war-mongers, victims and bystanders'. Who are which at any given time is presumably variable. In any conflict between the Superpowers of the other two blocs, the countries of the Third World, it might be assumed, would wish to be among the bystanders. However, as the Malay proverb puts it. 'When two elephants fight, it is the grass in between that gets trampled on.' Thus, the Bandung Conference notwithstanding, the leaders of the Third World countries have tried as far as they are able not to place their countries in the position of the 'grass in between'. This has often necessitated leaning towards one or the other of the two blocs. None has successfully played the one off against the other; even

Gamal Abdul Nasser, who was the best practitioner of this game, did not succeed entirely, as his successors discovered.[2] This necessity coupled with the need of military equipment for national defence and emotional post-colonial links led to the development of spheres of influence, with Third World countries lining up with one elephant, or, if the post-independence rulers were of a different colour (e.g. North Vietnam), with the other.[3]

The Third World countries also have the problem of 'Finlandization' to consider. This is where a country retains a high degree of autonomy and territorial independence from a very powerful neighbour, but in order not to invite trouble the country's foreign policy etc. is conducted with reference to that of its neighbour. Finland post-1945 is taken as the classic example of this situation; Bangladesh, Sikkim, Bhutan, Nepal might be others. Such political realities place certain constraints on political decisions for science policy and development on such matters as free access to foreign resources, the development of certain types of programmes, or the training of personnel. The danger of Finlandization, therefore, is that the country may slip into a surrogate or satellite status and eventually be worse off than it was as a colony.

With the increases in crude oil prices, the traditional gap between the 'too much' and the 'too little' has been bridged by the member nations of the Organization of Petroleum Exporting Countries (OPEC). However, bridging the gap has only created a new division between developing countries with oil, and those without.

Fanon's picture of a poor Third World now needs modification. One could divide the world into three *blocs*: developed countries, developing countries, and OPEC. One could further subdivide the developing countries into Underdeveloped Countries (UDCs) and Less Developed Countries (LDCs); the difference being that while the LDCs possess some capital and indigenous technology, the UDCs do not.

From the Muslim point of view, the world can only be divided into two categories: Muslim and non-Muslim. We will call the non-Muslim world 'Occident' and the Muslim world as a whole will be referred to as the *Ummah*. Perhaps we should explain our terminology in some detail.

The Concept of Occident

The Free World or the capitalist bloc is often referred to as 'the West'. An unfortunate ambiguity creeps into the use of its logical counterpart 'the East'; it may be that its users intend this duality of meanings. This confusion, however, is unnecessary and should be eradicated. The use of the West to refer to the Allied forces on that side of the line dividing

West and Middle Germany at the end of the Second World War naturally led to the use of the East to refer to those forces on the other side, normally the Communists. The alternative use of the West has been to indicate the culture and place of origin (mainly Western Europe) of the imperial powers of the recent years and the counterpart East to indicate the colonial peoples and their culture. The natural tendency, to identify the 'Communist East' with the previously colonial East was strengthened by the Bandung Conference in 1955.

However, to the Muslims there is really no difference in the cultural and territorial origins of the capitalist West and the communist East. Marx, working in the epistemology of European Jewry and in the milieu of nineteenth-century German culture and philosophy, was advocating solutions to the contemporary problems of nineteenth-century Europe. His comments on India and 'Oriental Despotism' verge on the inane. Engels was no different. The communist ideologues, Lenin, Trotsky and Stalin, were concerned with Marxist revolution in Russia and the consolidation of such revolutionary gains. The foreign policy of post-revolution Russia differed from that of pre-revolution Russia mainly in the motivation ethic. Indeed the aggressive expansionism of the 'international class struggle' made it more aggressive.[4] It was communist Russia which completed what the Czars had failed to do, and conquered the 'East' of the independent, Muslim and the Central Asian nations and later extended the western borders of Russia's empire to the River Elbe.

The Three Power Blocs

OPEC	Developing Countries	Industrialized
Algeria	Argentina	Australia
Saudi Arabia	Brazil	Canada
Indonesia	Cameroons	EEC
Iraq	Egypt	Spain
Iran	India	US
Nigeria	Jamaica	Japan
Venezuela	Mexico	Sweden
	Pakistan	Switzerland
	Peru	
	Yugoslavia	
	Zaire	
	Zambia	

Culturally and historically, Russia and all of the western parts of her empire belong to Christian Europe. The initial formation of European attitudes towards the Muslim 'East' was a product of the prolonged hostility of the Crusades, deriving from the loss of Christian-European control of the Mediterranean. Much of the earlier aggressiveness of the Portuguese and Spaniards and later of French, Italian and their northern co-religionists of the Netherlands and Britain can be traced to the Holy-War-Against-Islam mentality. This same mentality provided much of the legitimization for issues as diverse as colonial educational policies and the slave trade. Even when the role of missionaries and committed Christians decreased and the empires were dismantled, these attitudes remained. The old enemy has been largely forgotten; but the attitudes have become institutionalized. It will take a fundamental restructuring of attitudes and beliefs before the Christian–Islam conflict is resolved. This basic conflict was responsible for shaping many imperial policies and postures towards the conquered states for the majority of these states were Muslim.

This then is the basis of our concept of the Occident. We feel that although the attitudes and policies engendered by the Christian-Islam conflict have lost their original religious character, they regrettably remain. Their cultural source is Christianity and the recently discovered 'Judaeo-Christian' heritage. As Omar Austin has said,

> The era in which we are now living and which is slowly infringing more and more on Muslim society is a civilisation which was born, as we believe, of a very intense tension in what was originally Christian civilisation. As it appears to me there has always been a very important split in the Christian personality, as it were, throughout the ages during which Christianity dominated the West, or Christian Europe. There has been an important dichotomy between the sacred and the secular, the spirit and the flesh, between the state and the church. This is the split I mean. And I believe that what happened was that the tension between these two things between that body of European heritage which came from Judaeo-Christian sources finally split asunder European man. The European men opted for the secular rather than the sacred. And I think, the civilisation we are now living in is the result of that divorce in the western spirit between the sacred and the secular.[5]

It is this cultural source which has produced Marxism and capitalism as economic theories and parliamentary, congressional, and Soviet as governmental styles. This same source exported these ideas in the past and continues to export them now. Thus, this Occident is no longer restricted to Europe and 'Outremer' but has its 'Outremers' throughout the world. Anything which belongs to the Occident, whether found in Europe or in Asia, is Occidental. Any oriental who aspires to what is Occidental or who has achieved his aspirations is thus either Occidentalizing or Occidentalized. The term embraces Europe and its lifestyle, wherever it be.

The Muslim Ummah

In the Muslim world too many countries—for example Pakistan, Iran, Tunisia and Egypt—are governed by the Occidentalized elite—the result of an unfortunate legacy. Nevertheless, taken as a whole, the Muslim world is an ideological community. By the Muslim world we do not mean a particular geographical area, somewhat like the Third World of Fanon, but the International Brotherhood of Islam which manifests itself in the concept of *Ummah*. The Muslim *Ummah* is a supranational community with a common culture, and jurisprudence, and a certain self-consciousness—a consciousness of belonging together.

The Muslim *Ummah* does not consist of people who are of the same race, or speak the same language or have the same historical background. Rather the *Ummah* is a community of people with diverse backgrounds, different languages and a multitude of races. Yet all are bound together by the cardinal belief of the *Kalimah: laa ilaaha illa Allah, Muhammadu - Rasool Allah* —there is no God but Allah, Muhammad is his Messenger. It is this *shahadah* (witness) that has created the consciousness of belonging together. Every Muslim and every Muslim society—whether it forms a majority within a nation-state or a minority—has this consciousness. As such, the Muslims, as individuals as well as political units of the *Ummah*, are distinctly different from all other people and states; they have something in common which is exclusive to Muslims. They all belong to the *Ummah* wherever they may be, enclosed in whatever political boundary.

From the seventh to the twentieth century, Arabs, Turks, Iranians, Afghans, Indians, Indonesians, Berbers, Chinese, Europeans and Africans had one citizenship. They were all considered Muslim citizens, irrespective of their origin or the form of local government under which they lived. They had one overriding loyalty and one permanent devotion to the Muslim nation, since in spite of the breakup of the Caliphate into several states, Muslims continued to live under the

Sharia (Islamic law) from which all citizens derive equal rights and duties applicable wherever they happened to be.[6]

At present, the Muslim *Ummah* is split into two broad groups:

1.   Areas where Muslims constitute a majority, making some 46 nation-states containing 600 million Muslims. Many of those states are ruled by Occidental elites and in some like Albania (where Muslims are 65 per cent of the total population) members of the minority occidental groups are in complete power. This area constitutes approximately one-third of the total land area of the globe and forms a vital strategic belt between the free world and the communist bloc. We will refer to these 'nation-states' as 'Muslim countries'.

2.   Areas where Muslims constitute the minority. This involves some 300 million Muslims or a third of the total world Muslim population. The plight of the Muslim minorities of the Philippines and Ethiopia is well known. The treatment of Muslims in India, China and the Soviet Union is perhaps less common knowledge. But wherever they may be, whether persecuted, or contributing vigorously to the host community, the Muslim minorities have a strong sense of belonging to the *Ummah* as a whole. However, this study focuses on the Muslim countries where the community is dominant.

The present fragmented state of the *Ummah* has led to the universal desire of Muslims to promote 'Islamic solidarity'. This desire manifests itself in the mushrooming of a number of international organizations and conferences such as Islamic Secretariat, the Islamic Foreign Ministers' Conference, the Islamic Summits—all devoted to the mobilization of the intellectual, material and spiritual resources of the *Ummah*. The Muslim Institute has suggested a framework within which the goal of Islamic solidarity may be pursued:

1.   The raising of the Muslim's consciousness of belonging to the *Ummah* to a much higher level, so that parochial or other lower-level loyalties may ultimately disappear;

2.   The attainment of a much higher level of commitment of Muslim individuals and societies to the cultural and civilizational values and goals of Islam and a corresponding reduction of alien, mainly Occidental, civilizational influences;

3.   Commitment of new institutions, such as the Islamic Secretariat and the Rabita al-Alam al Islami, as positive instruments for

the promotion of the needs and interests of the *Ummah* as a whole;

4. The creation of and commitment to new institutions to undertake specialized functions for the *Ummah* as a whole and across 'national', ethnic, linguistic and other boundaries that now divide the *Ummah*,

5. Total acceptance of demands based on total commitment to Islam, including the willingness to sacrifice 'career', property, social relationships or even life for the collective good of the *Ummah* and the pleasure of Allah.[7]

In discussing development policies within the Muslim world it is important to appreciate that quite wide differences exist in the degree of industrialization and in the level of science and technology within the *Ummah*. For example, Pakistan with a population of about 75 million has 8 universities, 6 advanced colleges, 34 research institutes, 31 learned societies, and 1 national academy of science. Saudi Arabia, with a population of only 9 million has 5 universities, 11 institutes of higher learning and 6 learned societies and research institutes. Mauritania, on the other hand, has only 1 college and 3 research institutes.[8]

Despite such wide differences in 'development', the Muslim countries have similar problems and often propose similar solutions. Five objectives are typically advanced:

(1)   increase in the growth rate (Net National Income);
(2)   industrialization;
(3)   full employment;
(4)   more equal distribution of income, and
(5)   a favourable balance of payments.[9]

The last four are presented in varying orders of importance. In the following chapters we shall consider how these objectives relate to science, technology and development on the one hand, and the cultural and civilization values of Islam on the other.

NOTES

1. See Frantz Fanon, *Wretched of the Earth*, Macgibbon and Kee (London, 1965); R. Jenkins, *Exploration* (Paladin, London 1971); and R. I. Rhodes, *Imperialism and Underdevelopment* (Monthly Review Press, New York, 1970).
2. See G.A.Nasser, *The Philosophy of the Revolution* (Smith Keynes and Marshall, New York, 1959) and M. Copeland, *The Game of Nations* (Allen and Unwin,

London, 1969).

3. See D. J. Duncanson, *Government and Revolution in Vietnam* (Oxford University Press, London, 1974).

4. See J. F. C. Fuller, *The Conduct of War* (Allen & Unwin, London, 1961).

5. Omar Austin, 'Islam in the Modern World', *The Muslim 12*, 358-60 (Feb./Mar. 1975).

6. A. R. Azzam in *Egyptian Political Review*, July-August 1960, p. 34, quoted by L. A. Sherwani, *Studies in the Commonwealth of Muslim Countries* (Ummah, Karachi, undated).

7. The Muslim Institute, 'The Fourteenth Century of the Hijra.' A memorandum submitted to the Islamic Secretariat, Jeddah, 1975.

8. From *World of Learning*, 1974/75 (Europa, London, 1974).

9. These objectives form the basis of all 'development plans'. See for example the five-year development plans of Pakistan, Turkey, Indonesia and Egypt.

2 A MUSLIM VIEW OF SCIENCE

For decades scientists have enjoyed a privileged position in society. One price of this privilege has been the preservation of a dichotomy between science and non-scientific subjects. Snow[1] and Bronowski[2] have made brave attempts to overcome this dichotomy and close the gap between science and the humanities by focusing our attention on the aesthetic similarity between scientific and artistic creativity. However, the gap, if anything, has widened. For some scientists the benefits have been obvious. Not only have they been generously supported, but also their work had a chance to develop in a relatively isolated, detached environment, free from the morass of unsolved social problems. Protected from external pressure these branches of natural sciences have developed a level of skill and sophistication that is unique in our time.

Today, however, science has become so important, so powerful, as well as expensive that it is no longer possible for society to support it in the abstract. The very success of science has led it into politics: major policy decisions are more and more subject to public criticism, and 'science policy' must now be discussed in the open.[3]

This makes it all the more important that the nature of science is understood; how does it differ from other human activities?

In fact, science can be considered to be a set of human activities although many will disagree with this definition. For some science is simply a method, an objective methodology for establishing verifiable facts. For others, science is the coherent, growing body of public knowledge that has resulted from the cumulative application of this methodology. We consider science to be a complex combination of all three partial views. But more than that: we consider all aspects of science to be value-orientated and science as a whole to be a cultural activity, an activity that is shaped by the world-view of the actor.

Scientific Rationality

The most recent defence of science as 'objective truth', the neo-Apollonian ethic,[4] has been made by Wineberg who argues that once a scientist decides to enter into a contract with nature for the sole purpose of discovering her laws, he becomes nature's pupil.[5] The decision

21

in itself is a value judgement. However, having made his contract the scientist cannot deny the evidence provided by nature which indicates that these laws are impersonal and value-free. Wineberg's conclusions are, of course, implicit in the epistemology of the neo-Apollonian, rationalist ethics: and the support for science as value-free and intrinsically valuable human activity becomes an act of faith. The conviction that other dedicated individuals share that faith provides an important element of cohesion for the community of scientists.

The assumption here is that the set of values that guide the activity of scientists, the scientific technique of discovery, is value-free, and this gives science its pre-eminence. Science is claimed to have a special form of intellectual and moral integrity which makes it superior.

This is not to say that the neo-Apollonians do not question contemporary ethics but simply that they see science itself as the saviour: 'It is time that science, having destroyed the religious basis for morality, accepted the obligations to provide a new rational basis for human behaviour—a code of ethics concerned with man's needs on earth, not his rewards in heaven.'[6]

After a long and elegant exposition of the problems created by science and the lack of purpose in modern living, Harvey Brooks concludes his paper 'Can Science be Redirected?' with the following words:

> What I would like to suggest in conclusion is that science may to some extent begin to supply such a purpose. Material equity may simply never be enough to motivate a society, especially when its achievement may involve real sacrifices on the part of a significant segment of the population. Only goals in which all can share can energize a society. In many ways the space programme did succeed in doing just this briefly during the 1960s, even though the achievement of the goal turned sour in the end because of its apparent conflict with social equity programmes, and our realisation of how far we were from achieving the proclaimed goals of these programmes. In the past this kind of unified goal has only been achieved through war or national rivalry. The challenge of the future is to find other goals which are not structured around national or group rivalries, whose achievement is not at somebody else's expense, but in which all citizens can take pride. Science and technology may provide such goals.[7]

Can there be a better defence of science and technology than that? The argument, however, is not as sophisticated as it appears. It rests on four basic rationalist assumptions:

(1) that physical sciences provide the paradigms of objective knowledge;

(2) science is value-neutral;

(3) ethical and other normative 'judgements' are only expressions of emotions; and

(4) human behaviour is entirely determined by antecedent events so that it can be theoretically completely described or predicted by basic natural laws, established in accordance with the scientific method.

This epistemology equates science with truth; and therefore investigations must proceed, no matter what the consequences. This notion, as Roubiezek indicates, makes scientific knowledge an absolute value.[8] Once trapped in this epistemology, the neo-Apollonian finds it difficult to see beyond science: what could be more solid and concrete than an absolute frame of reference?

'Objective truth' in science is derived, of course, from the so-called 'scientific method', which involves human observation, experimentation, deductions, validation and subsequent attempted refutation — all leading to an accumulation of objective facts as a body of knowledge.

The currently accepted view of scientific objectivity treats observation as a direct sensory experience — touch, smell, colour, taste, and the like.[9] Scientific positivism in fact regards these experiences as fundamental to scientific method. However, a man's emotional condition at a particular moment may play as important a role in determining his conclusion as his sense of sight. In certain situations emotions will be a scientist's most powerful weapon. To ask him to ignore his feelings, emotions and sympathies is to ask him to deny his nature. This is neither possible nor desirable.

From Kant onwards, it has been understood that every set of scientific observations is dependent on some general theory for its collection. In order for any set of observations 'to test a theory A' we must presuppose some theory B to show how the observations relate to a test of theory A. Furthermore, for a set of observations to test a theory, it must in some fashion either be a function of or related to that theory. As such 'data' or collections of 'scientific facts' can never really be neutral, they must always express a viewpoint.[10] In *The Subjective Side of Science* Mitroff[11] has gathered some interesting evidence concerning the analysis of moon rocks by eminent scientists and how their findings supported their own pre-conceptions.

However, the scientific community would argue that its own internal processes of critical assessment weed out the arbitrary and subjective. The system of submitting papers to eminent referees ensures that shoddy additions to the edifice are discarded and only those that are safe to build on are retained. However, science like other areas of human activity, has its own politics. As for integrity, there are some who even claim that Newton himself, the archetype of scientific integrity, 'fudged' his results.[12] The sublime ideal of conventional scientific endeavour rests on an illusion of innocence, which can no longer be maintained.

Research itself is becoming political. Hilary and Steven Rose[13] have demonstrated that both the selection of areas for research as well as scientific concepts are politically and ideologically influenced in various degrees.[14] And as Jean-Jacques Solomon has said, 'the mystifying element in the ideology of science does not prevent it from leavening the behaviour of individuals who cling to the ideal of an intellectual community determined to question the meaning of the ends, served by the creations of the mind'.[15] The 'mystifying element' is, in fact, the leaven itself: the belief in the neo-Apollonian ideal. For as Solomon shows so convincingly, scientific power is necessarily subservient to political power, and modern science essentially aims at power.

It is not difficult to find examples of political dogma masquerading as scientific truth. Soviet genetics under Lysenko, Jungian psychoanalysis under the Nazis, and the current conflict over I.Q. and intelligence are but a few examples. It is naive to imagine that these are isolated events and that they are easy to detect and eliminate. They are easy to detect for they run counter to a stronger, well-established doctrine. But what if there is no other doctrine, because no one has thought of it yet; or those that have thought of it are not yet acceptable to the scientific (or political) community?

Partly because of its indubitable successes natural science has seemed for many decades to be the paradigm of genuine knowledge. In a variety of ways, the pursuit of science, or a popularised image of scientific work or scientific knowledge, has functioned as a religion in substitution for, or in opposition to, the accepted religion of a society. At the more practical level, the belief in the possibility of attaining some truth that will live on after one's personal death, has furnished a motive for selfless and dedicated work by many scientists; and some schoolmasters, forced by circumstances to transmit what is mainly a manipulative craft, derive self-respect and peace of mind from the conviction that they are imparting clear and distinct truths to their young charges. Since the pursuit of truth can appear to be the most noble and harmless of human activities, this conception of science helps to justify their work to scientists themselves; and it also helps them argue for public support, especially when public relations work is done on behalf of expensive experimental fields of research.[16]

To base a framework for ethics on the religion of science is, of course, to undermine the very basis of ethics. It is precisely the pursuit of such 'truths' which has created the present crisis in science and the consequent predicament of mankind. The importance of rationalist philosophy is not so much theoretical as pragmatic. In so far as one accepts the theory of determinism, one is able to construct a theoretical justification for a 'value-free science' and for ignoring ethical questions.

The real problem of the determinist thesis, however, is not its potential for misapplication, not even internal inconsistencies. Rather, it is its sterility with regards to the most basic concerns of ethics — that of personal decision making. Once a person has made a decision, the determinist can (theoretically, at least) provide a description of the factors which caused that person to make that decision. Likewise, the determinist claims to be able (theoretically again) to tell us how to cause another person to make a specified decision. But the determinist can tell us nothing about how to go about making the *right* decision, or how to determine what the right decision might be. And...if modern science and technology have done anything, they have created an increasing number of increasingly difficult decisions which must be made...[17]

Clearly, the positivist, rationalist account of ethics overreaches itself, however reasonable it may seem as an attack on institutionalized Christianity (the previous paradigm of intellectual respectability). To defend it is to put much more than just an academic epistemology at stake. It is the defence of rationalist occidental civilization against all its enemies. Unfortunately, this conflict between institutionalized Christianity and scientific rationality has come to be known as the war between science and religion. Science's temporary gain has been taken as the complete demolition of religion. However, institutionalized Christianity is not the only representative of religion. In fact, it does not even represent Christianity itself. There are many more enlightened religious outlooks and systems of belief which encourage learning, have their own traditions of science and with which modern science has never competed. And there are some world-views that do not even consider science as a real threat let alone a contender.

Scientific Mysticism

Recently, the position of the neo-Apollonians, the proponents of the traditional scientific ethics of rationality and positivism, has come under heavy attack for placing mankind in its present predicament. The current problems of environmental destruction, alienation between man and man, and between man and nature, and the crisis of ideology in science itself, are blamed on the conventional rationalist ethic. This attack is spearheaded by the neo-Dionysians, or the 'scientific mystics'.

The point of departure for the neo-Dionysians is that scientific rationality is *the* primary dehumanizing influence in contemporary society. There is therefore no question of science providing 'a new and rational basis for human behaviour'. According to the neo-Dionysians, scientific rationality must be replaced by a new ethics, which will legitimize sound ecological policy and will propagate a highly personal moral outlook. Thus Everett Mendelsohn asks:

> Can we create a new 'way of knowing'—a new epistemology fit to deal with the problems generated by a science engaged in a high industrial, a high-technology culture, and to deal with the connected crises that have emerged for all the human sciences? What shape would such a way of knowing have? That science, as a way of knowing and acting, needs reform at its very roots is something I am almost taking for granted.
>
> The signs are all around us and to pretend that all we have to do is to 'dig in' and weather another temporary storm bespeaks a kind of blindness. For one, just look at the relationship of science to war. Certainly one of the most destructive and tragic consequences of science's contact with secular authority has been its willingness to pliantly serve the authority masked by its claim of neutrality. Science, both natural and social, has become an integral part of a system of violence, a war-making system. Repeatedly, the sciences natural and social have claimed that their knowledge is neutral, that it is usable for either good or evil, but itself has no stricture in its development against doing violence. To have no stricture against violence is really to say that knowledge and technique can be bought

to do violence. The choice for violence by default is no less destructive than if consciously taken; it turns out also to be of course normatively loaded.[18]

Ravetz is much more open in advocating mysticism:

> Ah, so we have come to mysticism at last. Very fine, but what does this have to do with redirecting science? Surely it will require a political and institutional struggle for that; some scattered visionaries are quite irrelevant to the task. The only reason for starting here, with oneself, is that that is the only place one can start. Also, with this starting place, social awareness coupled with spiritual realization, a change is less likely to become one of those grotesque mockeries of ideals, in which people are forced to be free in some way that is decreed by the radical reformer. Such solutions only mask problems; they do not transform them. May I remind you again that no secular power-grouping or cause has stood back from the rush to nuclear annihilation; on this decisive issue they differ only in degree, not in kind. I think it is time to see whether a new conception of 'problem' isn't what we need.[19]

In short, the dominance of science must be replaced by a transcendent cognitive mode, which effectively denies the notion of value-free science.

Perhaps the best-known contemporary examples of the neo-Dionysian genre are the works of Roszak[20] and Reich.[21] Roszak, for example, sees modern science in a terrifying mould:

> The creature I name wears the face of despair; its lineaments are those of spiritual desperation; in its bleak features, scientists will see none of their own exhilaration and buoyant morale. They forget with what high hopes and dizzy fascination Victor Frankenstein pursued his research. He too undertook the adventure of discovery with feverish delight, intending to invent a new and superior race of beings, creatures of majesty and angelic beauty. It was only when his work was done and he stepped back to view it as a whole that its true — and terrifying — character appeared.[22]

He goes on to present gnosis as the legitimate substitute for scientific knowledge and quotes Paul Tillich describing gnosis as 'Knowledge by participation...as intimate as the relation between husband and wife'. Gnosis 'is not the knowledge resulting from analytic and synthetic research. It is the knowledge of union and salvation, existential knowledge in contrast to scientific knowledge.'

Similar criticisms of science, and echoes of mysticism, can be heard in the works of Illich,[23] Nasr,[24] Davenport,[25] Goodman[26] and Teilhard de Chardin.[27] All these scholars advocate a very personal morality; the object is to achieve salvation. There is a danger, however, that ethics of a personal and mystical nature may be seen as escapism.

'Salvation' is not enough, we also need an enrichment of life. Any operational ethic must give concrete shape to the ultimate spiritual and physical values in everyday activities. The ideal must be translated into a reality, and not simply in the future in the heavenly kingdom.

It is not surprising that science has brought occidental civilization back to age-old problems. Science, as we know it today, is a product of the occidental civilization, an embodiment of its culture and values.[28] The roots of modern science lie not just in the industrial revolution but in the Age of Enlightenment. Hear the *Philosophes* — the intellectuals who conceived and perfected the Enlightenment. What were the underlying assumptions and implications of science? What was science trying to do? Was it trying to enlarge the bounds of human empire, or to make all things possible, so as to give man mastery over nature, or was it trying to subdue nature to the human will for the benefit of humanity? In recent years science and the 'ethic of knowledge' has even been seen as incoherent

Lynn White has pointed out that the assumptions of modern science are those of medieval Christianity.[30] Rationalist science is thus based on the world-view of Christianity. So, if you change these basic assumptions to introduce a new world-view, you would have a different science. Consider the outstanding changes that could take place if the research workers of MIT were replaced by Taoist monks. If modern science had matured under Islam, it would have had an entirely different entity.

Science is a cultural phenomenon. Every culture has a view of the natural world, of society and of knowledge — be it conscious, or unconscious, well-articulated or incoherent. As all cultures consider the basic problems of man and exhibit rationality to some degree, they have some kind of science. It would seem that occidental culture contains the most rational of sciences. But other cultures too have had their sciences at the zenith of their civilization.

Modern science is the embodiment of occidental actions and culture, as is evident from its assumptions about the relationship between man and nature, universe, time and space. Islam makes very different assumptions about these issues and therefore an Islamic method of knowing, a science based on Islamic assumptions, would be an entirely different proposition.

Islam and Scientific Knowledge

Scientific curiosity and systematic investigation have been outstanding features of the civilization of Islam. This is not surprising for Islam is a rational religion; but not rationalistic. It has developed a sophisticated awareness of the central place of reason in religious traditions and in maintaining a critical attitude towards scholarship. Islam not only respects and encourages learning but includes within its framework rational methods of inquiry. As such, Islam has produced not merely some 'great scientists' but an entire tradition of science—a tradition which integrates scientific objectivity within the Islamic world-view.[31] The science of *Hadith (Ilm al-Hadith)*, which forms the basis of the moral and legal codes of Islam and provides a sophisticated methodology for the criticism of writing *tafsir*, commentary on the verses of the Qur'an, has also evolved a sophisticated methodology as well as a tradition of scholarship. Together the Qur'an and the *Hadith* have been the basis of all scientific activity in the history of Islam. The methods developed for the criticism of *Hadith* and the techniques for the scholarship of *tafsir* both influenced the humanities as well as branches of natural science.

Science developed a unique mould under Islam and flourished. Scholars in the Latin cultures of Northern Europe literally sat at the feet of Muslim scientists, in Spain and various Mediterranean centres, learning the rudiments of science and other aspects of the Islamic achievement. Only in the sixteenth century did European science and technology equal the best of Islam.

The tradition of science and scholarship developed by Muslims is indeed unique; but its uniqueness lies not only in its methodology but also in its epistemology.

The epistemology of Islam contains a holistic concept of knowledge. Here there is no divorce between knowledge and values. Knowledge is related to its social function and is considered to be an important characteristic of our being. As such there is a unity between man and his knowledge. There are no specialized, value-free 'bits' of information for specialized tasks. There is no dehumanization, isolation, or alienation.

Knowledge, in the epistemology of Islam, is like a tree and various sciences are so many branches of this single tree, which grows and issues leaves and fruit in conformity with the nature of the tree itself. However, just as the branches of a tree do not continue to grow indefinitely, so a discipline is not to be pursued beyond a certain limit. The pursuit of a particular branch of knowledge beyond its limits becomes a fruitless activity. If a branch of a tree grows indefinitely it would certainly end up destroying the harmony of the whole tree.

One of the best articulations of this epistemology is found in *The Book of Knowledge* of Imam Abu Hamid Muhammad Al-Ghazzali (1058-1111) (called Algazel by medieval scholars), a professor at the Nizamiyya Academy at Baghdad.[32] Al-Ghazzali analyzed knowledge on the basis of three criteria:

1. The Source.
   (a) Revealed knowledge: 'it is acquired from the Prophets and is not arrived at either by reason, like arithmetic, or by experimentation, like medicine, or by hearing, like languages.'
   (b) Non-revealed knowledge: primary sources of these sciences are reason, observation, experimentation, and acculturation.
2. The Level of Obligatoriness.
   (a) Individually Requisite Knowledge (*Fard-ulayn*): that is knowledge which is essential for an individual to survive, e.g. social ethics, morality, civil law.
   (b) Socially Requisite Knowledge (*Fard-ul Kifayah*): What is essential for the survival of the whole community, e.g. agriculture, medicine, architecture, engineering.
3. The Social Function.
   (a) Praiseworthy sciences: these are useful and indispensable sciences 'on whose knowledge the activities of this life depend....'
   (b) Blameworthy sciences: these would include astrology, magic, certain types of war sciences, genetic engineering, aversion therapy, the scientific study of torture, etc.

In this framework, science and the humanities exist not as 'two cultures' isolated from each other but as two pillars which derive their vital solidarity from the continuum of total human culture. Here knowledge is at once dynamic and static. There is gradual development of particular forms of science, while, at the same time, there is an awareness of the immortality of the principal knowledge derived from revelation. The Muslim framework of knowledge never loses sight of the revealed knowledge which provides the matrix for all human sciences.

The purpose of studying a subject in Islam is its importance to the community: its social relevance. There is no such idea as science for science's sake. Also rejected is the notion of a purely utilitarian science. The legitimation for the study of a science is found in the Qur'an where man is commanded to contemplate the heavens and the earth and all that is enclosed within. If you utter and believe 'Allah is Omniscient' then you have no justification for ignorance.

Within this circumference, Islam emphasises the importance of pure knowledge and encourages the pursuit of knowledge for the perfection of man.[33] What could be more useful to man than knowledge which is an adornment for his soul and the means for his attainment of perfection? However, the pleasure of attaining knowledge must combine with its necessity and social function. Knowledge, far from being enjoyed as an end in itself, must be instrumental to the attainment of some higher goal. All science whether its primary source is Revelation or scientific inquiry, can become 'blameworthy' if it loses sight of its ultimate goal.

How does a science become blameworthy? When it 'leads to harm'; when it creates undesirable social effects; when it tends to such a level of abstraction that it leads to the estrangement of man from his world and from his fellow man; when it leads to confusion not enlightenment.[34] The epistemology of Islam requires a methodology which takes account of the Inner Experience of Man as well as sensory perception, experimentation, deduction and induction. The experience of man as a complete being encompasses not simply physical and sensory stimuli but also intellectual intuition and psychic processes. Isolating the physical from the inner experience leads man towards depersonalization and alienation, and ultimately self-destruction. For the full man *all experiences are real*, as real as gravity, and therefore worthy of evaluation and investigation. To exclude any one of them is to exclude reality itself. Islam urges its followers to approach reality as a whole. A precise statement of this sanction is the basic negation affirmation phrase of Islam: There is no reality but the Reality (*La ilaha illa-Llah*). All other 'realities' are purely relative to and dependent upon the Reality. All cosmic determinations, whether formal or formless, subtle or gross, are nothing but indications of the Reality from which they stem by a process of creation or self-manifestation. For a complete vision of the Reality, sense-perception must be supplemented by the perception of what the Qur'an describes as *fu'ad* or *qalb*, or (in the equivalent English) the heart or inner eye. The inner eye is a kind of inner intuition or insight which, in the beautiful words of Mawlana Jalaluddin

Rumi, 'feeds on the rays of the sun and brings us into contact with aspects of Reality other than those open to analytical reasoning and sense-perception.' Indeed, it is something that 'sees'; and its 'reports', properly interpreted, are never false. This is not to say that the Inner Eye is a mysterious special faculty; it is, rather, a mode of dealing with Reality in which sensation, in the physiological sense of the word, does not play any part. Yet, the vista of experience thus opened to us is as real and concrete as any physical experience. To describe it as 'spiritual', 'mystical' or even 'supernatural' does not detract from its value as experience.

The occidental scientist is trained to evaluate nature in precise quantifiable terms. His knowledge must be verified through controlled physical experimentation and should be based upon laws or paradigms which pertain to its particular nature. Yet if his scientific mandates dismiss 'metaphysics' as a valid field of academic investigation, the scientist is in fact branding a large body of human knowledge, experience and intuition as unworthy of serious discussion. The alternative presented by modern occidental science is isolation, abstraction and estrangement. The critical question is, how do the social responsibilities of Muslim scientists compare with those of the neo-Apollonians and neo-Dionysions?

The Social Responsibilities of the Scientist

The scientist has three responsibilities: (1) to himself, for making the best of his life; (2) to the society and the environment; and (3) he has a responsibility to those inner feelings that determine for him that certain things are inherently important and valuable.

From the rationalist, neo-Apollonian point of view, responsibility towards the self simply means pursuing one's career. In his second responsibility, the rationalist insists that the society and the environment must change to suit his outlook; the external forces must adapt themselves to suit science so that 'objectivity' and 'progress' may be achieved. The third responsibility is often beyond him. If it is accepted at all, it is regarded generally as imposed from without in the form of laws to which the scientist is subjected without any choice. Even religious injunctions and prohibitions are regarded as something outside him which he follows merely to conform to the behaviour of his particular scientific community.

Scientists' responsibilities viewed from the rationalist perspective usually take the form of utilitarianism: the greatest good for the greatest number, which necessarily implies the sacrifice of the 'good' of the some. The three responsibilities of the rationalist scientist are unconnected, but run parallel, running into infinity and oblivion.[35]

In contrast, the 'mystic scientists' view the three responsibilities of the scientists as a co-ordinated whole, largely focused on the society and its environment and, depending on the variety of mysticism (Sufi, Buddhist, Divine Light, Catholic), the Hereafter. Mysticism is a means by which one acquires control over one's self and realizes the importance of inner experience, sincerity and the constant presence of God in all one's acts and thoughts, seeking to love God more and more. However, for a mystic the mere affirmation of these beliefs has no value; he must aspire to assimilate it and feel it as reality. Although this is an attractive position there are certain reservations. Mysticism has not been a strong force in the social and religious life of occidental countries for some centuries. The potential still exists and were there a major crisis in high-technology civilization, mystical and enthusiastic religions would certainly erupt but, without the discipline of tradition, they could easily assume a grotesque and destructive character. In this sense the neo-Dionysian approach to science is an overreaction to the ills of contemporary scientific ethics.

Islam presents a balanced view between the neo-Dionysian and the neo-Apollonian. This is seen most clearly in the treatment of time. In the rationalist philosophy time is a linear progression: for a particular individual scientist, time ends with his life. Beyond his life there is no time, at least as far as his own individual identity is concerned. For the mystics, only some kind of belief in life after corporeal death can make life on earth meaningful. Islam synthesises the two views: this life is life in time, while the Hereafter is the life in eternity, where we are able to pass beyond the limits of space, time and causality. We must look upon life as a tapestry in which time and eternity are woven together. This brings unity in the life of the individual and moulds science and society into a co-ordinated whole where each contributes to the welfare of the other, and no individual sacrifices his own interests. This concept of life resolves the contradictory and conflicting claims of scientism, altruism and egoism.

When the three responsibilities of the scientist are viewed in the larger context of life both here and Hereafter, his responsibilities towards himself and his discipline will not only be concerned with the events of his temporal life but also with the invisible part of his existence. He will see that his own personal welfare is intimately bound with the growth of his personality in the eternity of the Hereafter, so that he will look upon his personality in terms of his own inner experience. All

changes in external environment, according to this view, will be consequent upon the changes within him, his inner 'conversion'. His attitude
towards society will have the same orientation. When it comes to the
third category of responsibilities he will see them as not something
separate from himself, imposed from outside, but as identical with
something he has within himself. Thus these three responsibilities will
appear as three different lines all converging towards the same moral
ideals.[36]

The search for 'individual salvation' can thus lead to societal uplift
and metaphysical progress. However, the moral as derived from the
'tapestry of life and eternity' are not to be personalized as, for example,
suggested by the neo-Dionysian philosophy, or institutionalised. They
are to be socialized. The socialization of moral laws requires a de-
individualized morality, a change in the objectives of moral life
from personal individual salvation into social well-being and social
harmony.[37]

NOTES

1. C. P. Snow, *The Two Cultures* (Scribner, New York, 1971).
2. J. Bronowski, *Science and Human Values* (Harper Torchbooks, New York, 1965).
3. 'It is characteristic of basic and strategic science that neither the devising of programmes of work nor the assignment of relative scientific priorities to each programme can be carried out by non-scientists. If wise decisions are to be made, programmes must be scrutinised and assessed by scientists of wide knowledge, much experience and a broad synoptic view of science. However, it is of the utmost importance that those making these judgements should be continually aware of national needs and objectives. Otherwise there will develop a dangerous and corrupting 'ivory towerism' which will also impede the most effective transfer of scientific ideas and discoveries to practical use.' Sir Frederick Dainton in a report on the future of the research system in *A Framework for Government Research and Development*, Comnd. 4814 (HMSO, 1971), p.13.
4. After Gerald Holton, 'On Being Caught Between Dionysius and Apollonius', *Daedalus*, Summer 1974.
5. Steven Wineberg, 'Reflection of a Working Scientist', *Daedalus*, Summer 1974.
6. B. M. Oliver, 'Towards a New Morality', *IEEE Spectrum* 9, 52 (January 1972).
7. Harvey Brooks, 'Can Science be Re-Directed?' Paper circulated at Colloqué organized by Conservatoire National des Arts et Métiers ('Peut-on rediriger la science?), Paris, 4-6 December 1975.
8. Paul Roubiczck, *Ethical Values in the Age of Science* (Cambridge University Press, London, 1969).
9. The most widely accepted notions of objectivity are those of Leibnitz and Locke. Leibnitz's notions range from, firstly, the use of explicit well-formed rules to deduce systematically the conclusions — hypotheses, theorems —

of formal axiomatic systems, to secondly, the investigation of formal systems for the confirmation and falsification of provisional scientific facts or hypotheses, to, thirdly, formal rules for the design conduct and evaluation of scientific experiments. Thus there is no single notion of objectivity within the various Leibnitzian notions. However, there is a common thread through all of them: whatever the particular notion the emphasis is on depersonalized, careful testing and scrutiny of scientific knowledge by impersonal, well-defined formal tests. Locke also emphasized the impersonal, clinical character of scientific objectivity.

10. The point that if all observation is theory laden — as it must be if theory provides the guidelines for determining what is relevant to the testing of the theory — then the explicit development of any one theory of any phenomena can be detrimental to the continued testing of that theory is forcefully made by Paul Feyerabend in *Against Method* (New Left Books, London,1975).

11. I. Mitroff, *The Subjective Side of Science. A Philosophical Inquiry into the Psychology of the Apollo Moon Scientists* (Elsevier, Amsterdam, 1974).

12. See S. Westfall, 'Newton and the Fudge Factor', *Science*, 179, 751-8.

13. H. and S. Rose, 'The Myth of Neutrality of Science' in W.Fuller (ed.), *The Social Impact of Modern Biology* (Routledge and Kegan Paul, London,1971); and 'The Incorporation of Science', in H. and S. Rose (eds.), *Ideology of/in The Natural Sciences* (Macmillan, London,1976).

14. Why do we know so much about the electron? Because electric power is a fundamental need of a mass production industry, and because electronic communications are the foundation of modern governments. A thorough understanding of the social possibilities of the electron ended the General Strike in Britain, raised Hitler to power in Germany and led eventually to the final downfall of his system after Hiroshima. Can we expect any modern political machine to ignore such social functions?

15. J. J. Solomon, translation, *Science and Politics* (Macmillan, London, 1973).

16. J. R. Ravetz, *Scientific Knowledge and its Social Problems*, Oxford University Press, London, 1971, p.20.

17. Robert J. Baum, 'A Philosophical/Historical Perspective on Contemporary Concerns and Trends in the Area of Science and Values', *Newsletter 9 of the Program of Public Conception of Science*, Harvard University, October 1974, p. 30.

18. Everett Mendelsohn, 'A Human Reconstruction of Science', *Boston University Journal*, Spring 1973.

19. J. R. Ravetz, 'What Can We Learn from the Freaks?' Paper presented at the CNAM Colloqué, 4-6 December 1975, Paris.

20. Theodore Roszak, *The Making of a Counter Culture* (Doubleday, New York, 1969) and *Where the Wasteland Ends* (Doubleday, New York, 1972).

21. Charles Reich, *The Greening of America* (Random House, New York, 1970).

22. Theodore Roszak, 'Science, Knowledge and Gnosis', *Daedalus*, Summer 1974.

23. Ivan Illich, *Celebration of Awareness* (1971); *Tools for Conviviality*, (1973); *Medical Nemesis* (1975); *Energy and Equity* (1975) — all published by Calder & Boyas, London.

24. Hossein Nasr, *The Encounter of Man and Nature* (Allen & Unwin, London, 1968).

25. W. H. Davenport, *The One Culture* (Pergamon, London, 1970).

26. Paul Goodman, 'Can Technology be Humane? *New York Review of Books* 13, 27-34 (November 1969).

27. Teilhard de Chardin, *The Phenomena of Man* (1955); *The Future of Man*, (1959); *Man's Place in Nature* (1966) — all published by Fontana, London.

28. See Ziauddin Sardar, 'The Quest for a New Science'. Paper presented at CNAM Colloqué, 4-6 December 1975, Paris; also Muslim Institute Papers No. 1, Open Press, Slough, 1976.

29. As, for example, by Jacques Monod in *Chance and Necessity*, (Fontana, London, 1974).

30. Lynn White, 'Historical Roads of our Ecological Crisis', *Science* 155, 1293 (10 March 1967).

31. See the excellent works of Hussain Nasr: *Science and Civilisation in Islam* (Harvard University Press, Cambridge, 1968); *Introduction to the Islamic Cosmological Doctrines* (Harvard University Press, Cambridge, 1964); and his more recent *Islamic Science* (World of the Islam Festival Co., London, 1976).

32. Al-Ghazzali, *The Book of Knowledge*, translated by Nabih Amin Faris, (Ashraf, Lahore, 1962).

33. To see this framework in operational form in the life of a Muslim scientist, see Ziauddin Sardar, 'Al-Biruni, 973-1048: Encyclopaedist, Scientist, Philosopher', *Quest: Journal of the City University*, Summer 1974, pp. 28-31.

34. Consider, for example, the current debate on genetic engineering. *New Scientist* asks 'The realisation that people may be genetically vulnerable to particular diseases has created an ethical dilemma. Bluntly, should workers be selected for their genetic "immunity" to hazards or should the work environment be made safe for all workers by eliminating hazards?' While this question is debated, the work continues as no one wishes to the left out of the Grand Prix. See *New Scientist* 71, 486 (1976).

35. 'Such, in outline, but even more purposeless, more void of meaning, is the world which Science presents for our belief. Amid such a world, if anywhere, our ideals henceforward must find a home. That Man is the product of causes which had no prevision of the end they were achieving; that his origin, his growth, his hopes and fears, his loves and his beliefs, are but the outcome of accidental collocations of atoms; that no fire, no heroism, no intensity of thought and feeling, can preserve an individual life beyond the grave; that all the labours of the ages, all the devotion, all the inspiration, all the noonday brightness of human genius are destined to extinction in the vast death of the solar system, and that the whole temple of Man's achievement must inevitably be buried beneath the debris of a universe in ruins — all these things, if not quite beyond dispute, are yet so nearly certain, that no philosophy which rejects them can hope to stand. Only within the scaffolding of these truths, only on the firm foundation of unyielding despair, can the soul's habitation henceforth be safely built.' Bertrand Russell, *Mysticism and Logic* (Allen & Unwin, London, 1910), p. 41.

36. For a slightly more detailed exposition see Bashire Ahmad Dar, *Quranic Ethics* (Institute of Islamic Culture, Lahore, 1960).

3 SCIENCE POLICY AND DEVELOPMENT

In the 1960s when development first became a catch-word in international politics it was synonymous with 'progress'. The countries of the Occident were considered to be 'developed' (industrially, economically, technologically, institutionally and often culturally). Others were considered to be progressing towards the goal of development on the occidental model. This is a very ethno-centric view of the world; a late manifestation of the social Darwinist ideas of the Victorian era, which produced such notions as the 'White Man's Burden', 'Manifest Destiny', pressures for 'Reforms' in the Ottoman Grand State, and the like. The norm of development was the Occident and in order to develop, or be developed, one had to occidentalize oneself. Implied in this was an abandonment of cultural and other legacies which inhibited progress towards this goal.[1]

The critical points in the development continuum are at one extremity developed and, at the other, undeveloped. In between there are two transitionary stages, which we have grouped together as 'developing'. Listed, this continuum would look like this:

1. Developed countries.
2. Developing countries:[2] (a) less developed; (b) underdeveloped.
3. Underdeveloped countries.

Scafeti[3] states that developing countries are conventionally characterized by comparison with already industrialized countries, regarding:

(1)   *per capita* income;
(2)   percentage of labour force in agriculture;
(3)   energy consumption;
(4)   literacy rate [sic];
(5)   productivity; and
(6)   general consumption.

It is highly unlikely that such a place as an 'Undeveloped Country' could be found, but it is not uncommon to find areas of underdeveloped land. As development is assumed to be connected with the process of industrialization, a place like the Sahara Desert is, by and large, under-developed although it is much used by groups of people such as the Touareg.

Within the Muslim world there is a great diversity in the economic, scientific and technological states of the countries, but one could easily place them all in the category of 'developing countries'. This would mask the 'near-developed' nature of less-developed countries such as Turkey, Pakistan or Egypt, and the near-underdeveloped ones such as Niger, Mali or Chad. In between these there are a number of less-developed countries like Iran and Malaysia rapidly approaching the same state as Egypt, Pakistan or Turkey. However, most Muslim countries would be classed as less developed.

What is Development?

To make sense of the concept of development, we must define what it means. Lucien Pye provides the best definition describing it as a multi-dimensional process of social change. It is not simply a case of installing a water-desalination plant in an arid region, but also a question of man-power to operate it, the disruption of local economies and habitats unused to abundant water, the problem of salt disposal and so on in-cluding the changes wrought in the educational system.

In contrast, Alpert's declared aim for development is 'simply to modernise the developing countries and to raise them to the level of the advanced industrialised nations'.[4] Here development has become a process of modernization; namely, a linear movement towards economic growth, industrialization, and dominion over nature and environment. The obvious inference from Alpert's statement is that advanced industrialized nations are modern and that that is a desirable thing to be, and that the developing countries are not modern and that is not at all a desirable thing to be. Alpert's choice of the word 'modern' is unfortunate. As in the title of Daniel Lerner's book *Passing of Tradi-tional Society: Modernising the Middle East*,[5] the word is value-laden and harks back to the social Darwinism of the nineteenth century.

We must reject the word 'modern' because it has no meaning in this context. People, and the societies in which they live, by virtue of their being alive today, are modern. Similarly, we cannot accept that words like 'medieval', and 'barbaric' can be applied to Muslim societies and cultures. The Muslim world had no 'middle ages', 'dark ages', 'rennais-sance', 'reformation', or 'enlightenment'. Modernity in such usage implies a movement towards perfection; a perfection which the advanced industrialized nations almost embody. We cannot accept this. To a Muslim the perfection of human society is a society of Islam, ruled, formed and guided by the *Shariah.*[6]

Although certain occidentals like Pearson[7] have recognised that the term 'development' itself is quite inadequate, it still persists. It has been redefined to exclude the more obvious cultural overtones but no matter how the coy academic placates his own susceptibilities the old development and its effects continue because people still aspire to develop *à l'occident.*[8, 9, 10]

The occidentalizing dynamic of development policies must therefore be stated. Any theory of development seems to imply acculturation as long as the development is international or intercultural. If it were possible to have an *intra*cultural theory of development then it would presumably be possible to eliminate acculturation as far as that involves alien cultures. Not that this is necessarily a bad thing, as cross-cultural stimuli often produce highly desirable results. But, it is not the effects of benevolent cross-cultural fertilization that cause concern. It is the cultural depradations and freebooting of occidentalism which have aroused this passion.

We would define development as a strategic compound of private and collective actions, with their intended and unintended consequences, through which a society moves from one state of organization, one system of ideas, beliefs and traditions, and one stock of equipment to another in the context of other societies which have followed or are following a similar (though far from identical) route with similar (though differing) hopes, aspirations and fears. Development is far from being a simple, anodyne, economic process of raising living standards or increasing the rate of growth.[11] As currently understood it is a cultural process which inevitably leads to occidentalization.[12]

Thus the emphasis from the experiences of 'developing' countries should be that development means a definite cultural shift, and that this is sometimes an intended strategic consequence and sometimes (indeed, often) an unintended consequence of some superficially attractive scheme embarked upon lightly, from the point of view of culture.

Goals of Science Policy

On a very simple level we can consider science policy to be the course adopted by governments to promote science and technology for development. However, all policy decisions are subjected to political criteria; and science policy is no exception. Indeed Leiserson has defined science policy as 'specifying the criteria for allocating by political decision the appropriate portion of national or world resources devoted to the growth and the direction of scientific knowledge and personnel'.[13]

One's outlook on science policy depends very much on how one conceives development. Since the 1950s the specification of criteria for science policy has been based on the ideology of the *technocratic* utopia; and science and technology are used as the main tools for achieving this neo-Apollonian vision.[14] Thus Muslim countries have generally directed their science policies towards building infrastructures for the occidentalized types of nation-states.

In general, science policy in the Muslim countries has been directed towards three goals:

1. Strengthening the military power and the political legitimacy of the state. As such a large segment of science policy is in fact 'defence policy'. Often greater credibility to the political decisions of the state is given by backing them with the opinions of 'scientific experts' and commissions.

2. Promoting the economic development of the state. This results from the need to ensure the development and therefore competitiveness of industry and of agriculture. In the pursuit of this goal, science policy and economic policy have often merged into a co-ordinated whole. For example, the establishment of certain research installations and universities are supported not because of the specific scientific results that may be produced, but more for the side effects these installations may have on economic development and promotion of indigenous technical capabilities and qualified manpower.

3. Finally, science policy may be directed towards social goals, for example, the reform of the education system, the establishment of an information and communication system, the increase of social communities. Here science policy diffuses into the domain of social policy.

The pursuit of the goals of science policy often leads to the establishment of either ministries of science and technologies or national research councils which advise the governments on science policy and development strategies. As the model of pursuit is that of the Occident, the entire organizational structure promotes only the occidentalized goals for science policy and patterns for development.[15]

Perception and Strategy in Development and Science Policy

From our point of view, as understood traditionally, science policy and development can be reduced to two components; one of perception and one of strategy. In so far as science policy involves what has come to be known as 'thinking about the future', it shares these two components with development. Thus as the traditional meaning of development implies a linear teleology or a continuum, and science policy requires a 'vision of the future', there are two standpoints from which these two elements can be viewed: that of the developed (viz. the Occident) and that of the developing (in our case, the Muslim *Ummah*).[16]

By and large a country is developed (or developing) because it is perceived as being so. The developing country perceives itself to be underdeveloped and determines to develop according to the agreed values. It seeks help to achieve this according to the model of occidental economic theory; hence the demands for industrialization and sophisticated prestige schemes (e.g. atomic energy or aerospace programmes). Thus we can say, with the late Peter Nettl, that

> ...development has become, at any rate conceptually, a universal priority, a self-generating aspiration resulting from status inequalities in the international system. We must not fail to distinguish between the universality of the pressures for development, and the problem of how societies respond to this pressure...It is the unity and universality of the pressures for development that have led people to see the world as divided into two categories of society, developed and less developed. This universalization, this coherence of categories, is new, as is the passionate involvement in the problem of many of the leaders in the developing countries themselves. But it does not follow that because the actors involved perceive the situation in this fashion, our analysis of the responses to the pressures must also be carried out in terms of these same concepts.[17]

The other component of development is strategy. This is an interesting word which can convey quasi-militaristic overtones. Indeed there is definitely an element of Grand Strategy about many aid programmes and about relations between developed and developing countries. In this sense developed countries may perceive some developing countries as a better strategic risk than others (say, for example, India rather than Pakistan, Iran rather than Turkey, Venezuela rather than Cuba). Development may constitute also an element *in* Grand Strategy. In this imperial sense, development may not be much (if at all) different from acculturation. (Take for example the policies pursued .by China in East Turkestan, India in Kashmir, Russia in the Ukraine, the Central Asian Republics or the Baltic States.) However, strategy need not necessarily be understood in this way. It has a wider application. In committing a country to pursuing the goals of development, strategic decisions are taken of which the consequences are not fully understood. Furthermore, committing a country to effective programmes of investment and industrialization with the attendant training programmes requires strategic planning. Similarly committing the developed countries to the support of developing countries requires parallel strategies. It would seem to be a truism that countries seeking to develop should make strategic choices for their *own* real best interests. It is a matter of regret that all too often this has not been the case, particularly with regard to science, technology, and industrialization in the Muslim world. In opposition, the developed countries have often followed very aggressive strategies in pursuing their interests so that the access to technology and development capital has been tightly controlled. This is a principal factor in the widening technological gap between developing and developed countries.

Is Development only Advancing Prosperity?

It is increasingly realized that the strategic effects of many development programmes have been disastrous. It is, of course, a pity that these effects were not anticipated before the programmes were initiated. However, as many of these effects have been cultural, and the by-products of the transfer of technology, in general the occidentalized elites which took the strategic decisions do not view such consequences too seriously. These elites are committed to an emotionally held assumption that 'West is best': so many a Muslim can wax almost poetic about the need for family planning, monogamy and the need for industrialization.

A principal cause of this state of affairs is the definition of development in economic terms; such a materialist view must militate against religious or cultural values. Thus, occidental development theories which have tried to avoid the crusading assumptions of occidentalism by defining themselves in economic terms are unavoidably atheistic, or at best merely secular. Inevitably this tends to undermine Muslim cultures, for most of the culture is religious in origin. If a society's culture is considered expendable in the course of development, the culture is doomed once the development strategy is decided. Various standard arguments are used to justify these strategies. Usually it is that development is necessary to raise the living standards of a country, but if it is necessary to raise the living standards of a country, is development the best or the only way of doing so? Are the policies being proposed the only ones available? The Indian writer S. A. Rao has proposed one strategy of development:

> The development of the Third World is primarily development of its people to enable them to raise their living standards. . .Most regions of the Third World possess certain traditional occupations which have come into existence largely on account of natural factors such as people's hereditary skill, living habits, availability of resources and demand for products. However, these are in a disorganised state and suffer from stagnation.[18]

'Development of the people' sounds rather ominous. Wholesale development along the lines proposed by Rao is very dangerous, if one values the cultural integrity of the society. The breakdown of traditional values is too high a cost to pay. To equate development with raising the living standard seems *prima facie* to be an eminently just and reasonable goal but it is a false assumption. Development is not a simple economic operation and in fact Rao admits as much:

> As the project [a dairy products co-operative in the Gujerat, India] depends on the participation of the whole community, its development along scientific lines cannot but have far-reaching influence not only on the living standards but also on the attitudes of the people. . .Another outcome of far-reaching significance has been the steady breakdown of caste and other social barriers which are traditional features of most Indian villages.[19]

If this is the effect of a dairy products co-operative what about a steel mill or a car assembly plant? It may be that the breakdown of caste is objectively a good thing, but normally the occidentalized elite charged with the strategic choice on behalf of the whole society is unable to make an objective analysis of the society's needs because the elite no longer shares the society's culture.

Strategies of change which respected the society's culture would result in piecemeal borrowing and adaptation of sciences and technologies. There would be more in the way of substituting trade for aid. The domestic organization and operation of development programmes would vary, from society to society, and Rao's self-help idea which might be applicable in Gujerat would not necessarily apply elsewhere.

What is the Role of Science Policy in Development?

One could say that the role of science policy in development ought to establish clear criteria derived from an objective analysis of the development needs of the country, for government and other institutions to use in their efforts to provide the resources, scientific knowledge and personnel to service those development needs. In this case what sort of development is necessary? In essence it is a strategy of equipping the country to use its natural resources to its own best advantage: improving its trading position in relation to those markets most likely to suit its requirement; improving the quality of life of its citizens; improving the governance and economy of the interior and so on. All this must be viewed simply in terms of the real interests of the country and not in terms of some distant occidental notion of what is desirable. What holds for science and technology also holds for education, industrial training programmes, communications, defence technology, the production and training of civil servants; these and other fields should not be occidentalized but domesticated.

People may be trained and educated in the Occident, and the products and the capital of the Occident may be used freely, as appropriate. We are not advocating a 'four legs good, two legs bad' *simplisme*. What we are advocating is that scientific communities, and anybody else concerned with science policy should be honest about the real needs of their society and about the most appropriate solutions to their problems viewed not necessarily in isolation from the rest of the world, but as a unique society which has problems of unique dimensions. We should learn from others, but there is no need for us to borrow their spectacles as well.

In Muslim societies this borrowing of other peoples' spectacles is a result of lack of self-confidence. If a society lacks confidence in its cultural sufficiency, then it will allow itself to be changed by the new technology.

We believe that recent social damage in many Muslim countries is only partly a product of colonialism; partly it is self-inflicted—the product of a general lack of self-confidence within these traditional societies. This phenomenon has been observed by Rahman, among others, in the behaviour of the Indian scientific community towards British scientific developments.[20] Without self-confidence there can be no self-control and traditional societies are exposed to the random social change brought about by technological innovations. But need all social change be undesirable? The answer is surely 'No'. Any society, as it passes through its history, becomes encumbered with cultural barnacles of one type or another and some form of periodical screening is desirable. Otherwise the society will become overwhelmed by its accretions as well as unresponsive, inflexible and moribund. Social changes induced by the stimuli of trade, technology, war and so on can be beneficial no matter how traumatic their occurrence: indeed the probability is that they will be beneficial. In accepting social change the essential guiding principle must be some type of cultural monitoring which preserves the identity of the society. There are three principles upon which science policies in a Muslim country should be based:[21]

1. Science policy should be directed towards counteracting the cultural and environmentally destructive effects of concentrated science and technology. Emphasis should be given to the creation of indigenous scientific and technological capacity; the import of foreign technology and consultants must be checked, and local technology and expertise must be promoted.

2. Science policy must be distinguished clearly from defence policy and must be subordinated to economic and social development; science and technology are not an end in themselves, but only a means for achieving broader, more enlightened goals.

3. The formulation of science policy should not ignore the context within which the economies of other Muslim countries operate.

The development of local technological skill and scientific expertise should be guided by a strategy of selective interdependence: the Muslim countries should provide backup for each other and promote the development of a pool of Muslim scientific and technical experts. Close coordination and linkage between Muslim countries will avoid the probability of developing isolated technologies that would be rendered useless.

In the Muslim world it should become an axiom that cultural strength must be a factor in a strategy for science policy and development. This should be a general rule and not restricted to science policy. Muslim countries should favour the general strengthening of indigenous cultures, and not view it with suspicion, for this will make for a more dynamic yet stable country. Stable countries make better trading partners and whereas trade benefits all, aid becomes a burden and embarrassment to all. Aid indeed is to be deprecated, as its history indicates that the position of aid donor is all too often culturally subversive.

A developing country should thus be encouraged to trade rather than hold out the bowl and become 'a nation of beggars'.

Problems for Science Policy in Muslim Countries

We have emphasized that the essential principle for specifying the criteria for science policy must be domesticity and not occidentalism. The preservation of a society's cultural integrity should be the guiding light of the governing and scientific communities in a country. We consider this to be true of developed as well as developing countries. There should be no cultural or scientific imperialism. Therefore, we have used the catchword 'domesticity' as opposed to occidentalism to identify this principle. The details of the criteria by which resources are allocated must be then developed from this foundation. For Muslim countries it will mean pursuing policies which take their societies away from occidentalism and towards Islam.

The question that now arises is: who are the people who should decide these criteria? Normally it will be elite groups in the society: politicians, civil servants, scientists and the military. But these are just the people who are most occidentalized.[22] Without a careful and thorough self-examination, it is most difficult to see how such people can preserve the cultural integrity of their society.

If, then, both politicians and scientists are culturally representative of the outside,[23] who shall speak for culture? The answer is the alternative elites of a society such as the traditionally educated scholars, the religious leaders and tribal chiefs. These people must be involved in the process of specifying the criteria for science policy, as was the practice in the Ottoman Grand State where the preservation of cultural integrity and autonomy were part of the political theory.[24] The creation and training of scientific personnel must also involve the alternative elites and must be based on the principle of domesticity as it is an extension of the science policy.

The gravest problem for science policy in developing countries lies in the provision of external resources. Neither the attitudes nor the orientations of the donors or the receivers are conducive to the preservation of traditional culture. We disagree very strongly with Rahman and others who say that a scientist can be occidentalized and yet be a good Muslim, Hindu or whatever. He can only be one of two things: a completely occidentalized scientist who retains some contact with his traditional religion or a Muslim who has some contact with occidental science. The degree of contact may vary; there can even be total contact. The crucial factor is attitude. Attitude is the framework of assumed values and norms with which he approaches his subject. If he accepts *in toto* the attitudes of the Occident he must be occidentalized. If his attitude is Muslim, then he cannot but see science according to the traditions of Muslim science which are, in fact, the true basis for modern science.

There are, of course, other factors and problems related to science policy in the Muslim world. However, the key issue is the need to preserve Muslim identity and Muslim culture.

NOTES

1. See Ziauddin Sardar and Dawud Rosser-Owen, 'Science Policy and Developing Countries', in I. Spiegal-Rosing and D. de Solla Price (eds.), *Science, Technology and Society* (Sage Publications, London, 1977), ch. 15.

2. Michael Graves compares developing countries to a 'multi-divisional company struggling to survive in a highly competitive market. It is a small company, a recent entrant to the market and its underlying position is weak. The narrow range of products it can produce efficiently are, in the main, in excess supply. Most of the other items in its product line can be made in greater volume, to higher standards and at lower prices by its advanced competitors. Its cash flow is unreliable, its debt-equity ratio is bad and getting worse, its top and middle management teams are good but small and there are few trained manpower reserves to call upon. The country's price-earning ratio is low and its shareholders are living on expectations.' *New Internationalist*, II, 2-21 (January 1974).

3. A. C. Scafeti, 'Implications of Agriculture and Industrial Development' (Systems Development Corporation, Santa Monica, California, 1969, unpublished.)

4. Paul Alpert, *Partnership or Confrontation? Poor Land and Rich* (Free Press, New York, 1973), p. xi.

5. Free Press, New York, 1958.

6. Literally 'The clear path to be followed': it is usually translated as 'Islamic Law' but *Shariah* is not 'law' in the narrow, occidental sense of the word. It comprises, without restriction, the whole kaleidoscope of human activities, political, social, legal, domestic and private life of those who profess Islam. The sources of the *Shariah* are the Qur'an and the *Sunnah*, which comprises what the Prophet Mohammad said, did and agreed to. See A. H. Qadri, *Islamic Jurisprudence in the Modern World* (Ashraf, Lahore, 1973) and Said Ramadan, *Islamic Law. Its Scope and Equity* (Islamic Law Centre, Geneva, 1970).

7. Lester Pearson, *Partners in Development* (Prager, New York, 1969).

8. The now common term 'Third World' is equally value-laden. It, too, is in the mainstream of European historical experience. The original term *Tiers Monde* used by Fanon is evocative of France at the time of the Revolution: a parallel of the unrepresented *Tiers Etat* (predominantly, let it be said, *bourgeois*). Indeed, one might say, with the failure of various UNCTADs and the conspicuous evidence of petro-dollars, that he had a point. Fanon, a product of the French *Mission Civilisatrice*, as has been pointed out by the late Malik Bennabi, was educated to accept the values and beliefs which are tolerated in Metropolitan France, but yet emotionally he is not French. Such an effect of education is typical of the era of colonialism which is by

no means over and which has acquired a new form as well as dynamism.

9. For a review of occidental ideas on development, see Sardar and Rosser-Owen,'Science Policy and Developing Countries.'

10. Consider the commentary and definition of development by Denis Goulet: development 'designates simultaneously two realities: a terminal *condition* and a *process* by which successive approximations to this allegedly desirable condition are made. Terminal state is not to be construed statically here, but merely as embodying levels of living more acceptable than earlier ones. I accept the postulate that new levels of material wealth may possibly be required to provide support for the attainment of certain values by individuals, groups and societies. "Development" as here employed thus covers the entire gamut of changes by which a social system, with optimal regard for the wishes of individuals and sub-systemic components of that system moves away from a condition of life widely perceived as unsatisfactory in some way toward some condition regarded as "humanly" better. These changes may be gradual or mutational. And if they are to qualify as "development" some degree of calculation must be present on the part of society's influential decision-makers regarding optimum speeds at which change ought to proceed and minimal costs be paid. It matters greatly, of course, which values ought to be fostered in the effort to obtain a "humanly better" life. In fact, everyone who defines development makes an explicit or implicit option regarding several basic values. Among these are the degree of mastery to be exercised by persons over things, the level of critical awareness to be judged desirable in different categories or human agents, the optimal sharing of power to make decisions, and the destruction of particularistic interests in favour of wider reciprocity and solidarity. Certain quantitative improvements (in food, in real income, in suitable educational services, in increased life expectancy, etc.), important elements of development are not subject to quantitative measurement. Diverse ideal models of the good society exist; attitudes vary toward material benefits and the relation of these to personal wisdom, enlightenment, harmony, and other values. Development must be gauged by the values a society itself, or some member thereof, deems to be requisite for its health and welfare. This evaluation requires careful attention to differential abilities to produce wealth but also to less visible capacities such as that of processing information and providing "meaning" to life itself. In this sense at least, development can be described as the "maturation", "humanisation", or "qualitative ascent" of human societies.'

Earlier in the book, Goulet equates development with power: '. . .those countries which have gained the victory over space and time have also won privileged access to the world's resources. This is why development ethics must cope with structures of differential wealth and power among societies'. Denis Goulet, *The Cruel Choice* (Atheneum, New York, 1973, pp. 19, 333.)

11. This also seems to be the reaction of Gunnar Myrdal and other structuralists and the propagator of the Latin American theories of under-development. For the latter, a theory of dependency seems to be a substitute for analysis: after all the trumpets of the exploitative nature of capitalism and communist imperialism have been blown, the only conclusion that the Latin American theorists can reach is that the occidentalization of traditional societies is desirable! See for example, O. Sunkel and P. Paz, *El Salidesarrollo Latino-curericono y la Teoria del Desarrolo* (Siglo xxl editores, Mexico, 1970); J. L. C. Cerrantes, *Superexplotacion Dependencia y Desarrolo* (Editorial Nuestro Tiempo, Mexico, 1970); and Gunnar Myrdal, *Asian Drama, An Enquiry into the Poverty of Nations* (Pantheon,New York, 1968).

12. See D. Apter, *Ideology and Discontent* (Free Press, New York, 1964); L. Binder, *The Ideological Revolution in the Middle East* (Wiley, New York, 1964) and E. Ginsberg (ed.), *Technology and Social Change* (Columbia University Press, New York, 1963).

13. A. Leiserson, 'The Politics of Science: Science Politics, Science Policy, Policy Science — the Whole Thing',*Polity*, 6 (1), Autumn 1973.

14. An excellent example of this is provided by R. S. Bathal, 'Science Policy in Developing Nations', *Nature* 232, 227-9 (1971).

15. According to Spacy, the aims of science policy are of three kinds; defence, economic and social. see J. Spacy, 'The Problem of Choice', in 'Problems of Science Policy' (OECD Seminar, Jovy-en-Josa's, France, 19-25 February 1967), p. 17.

16. J. P. Nettl, 'Strategies in the Study of Political Development', in Colin Leys (ed.), *Politics and Change in Developing Countries* (Cambridge University Press, Cambridge, 1969), p. 15.

17. S. R. Rao, 'An Example for the Third World', *New Scientist* 59, 451-2, (23 August 1973).

18. Ibid., p. 452.

19. See A. Rahman, *Anatomy of Science* (National Publishing House, Delhi, 1972) regarding the attitudes of the Indian scientific community, which has its widespread analogies around the world, e.g. F. R. Sugasti, 'A Framework for the Formulation of Technology Policies, A Case Study of ITINTEC in Peru' (International Forum of Technological Development, 1975, unpublished); A. B. Zahlan, 'Science in the Arab Middle East' (1967, unpublished).

20. See also F. R. Sugasti, 'A Framework for the Formulation and Implementation of Technical Policies

21. Cf. H. G. Johnson, *The Role of Military in Developing Countries* (Princeton University Press, 1962); H. E. Davis, *Government and Politics in Latin America* (Ronald Press, New York, 1958); M. Halpern, *The Politics of Social Change in Middle East and North Africa* (Princeton University Press, 1963); F. Heady, *Bureaucracies in Developing Countries: Internal Roles and External Assistances* (Indiana University Press, Bloomington, 1966); M. Janowitz, *The Military in the Political Development of New Nations* (University of Chicago Press, 1964); C. Leys, *Politics and Change in Developing Countries* (Oxford University Press, 1969).

22. Rahman, *Anatomy of Science*.

23. Cf. T. Z. Tunaya, *Islamicilik Cereyani* (Baha Matbaasi, Istanbul, 1962).

24. Rahman, *Anatomy of Science*.

4 CULTURAL AND ETHNIC DIMENSIONS OF DEVELOPMENT

We have used the term 'culture' on several occasions without actually specifically defining it. We have said that science is a cultural phenomenon. We have also said that the Muslim countries should favour the strengthening of their culture and that cultural strength must be a factor in any strategy for development. But what is culture; and more specifically, what is Islamic culture?

As far as the English language and anthropology are concerned the varieties of definition, and the ambiguity of the concept of culture is notorious. To a great many anthropologists culture is 'learned behaviour'. For others, it is not behaviour at all, but an abstraction from behaviour. To some stone axes and pottery bowls constitute culture, while no material object can be culture to others. Yet for still others culture exists only in the mind.[1]

In its anthropological sense, the origin of culture is recent. Before 1843, 'culture' meant to cultivate, and it continues to have the same meaning in 'horticulture' or 'to cultivate the soil'. This was the basis of the word 'cultured', meaning refinement of taste and manners. Cultivators, those who 'cultivated the soil', considered themselves a step above the non-cultivators, those who were hunters or fishermen. The German ethnographer Gustav E. Kleman in 1843 gave an anthropological meaning to the word, and the classic (and the most often quoted) definition was provided by E. B. Tylor in 1871 in the opening lines of his *Primitive Cultures*: 'Culture is that complex whole which includes knowledge, belief, art, morals, law, custom, and any other capabilities and habits acquired by man as a member of society.'

For Margaret Mead 'culture' is the term applied to the total shared, learned behaviour of a society or a subgroup, so we may speak of 'a culture', using the term for the whole, or of an item of behaviour as 'cultural', referring this item to the whole. The model situation on which the anthropological concept of culture is based is that of the total learned, shared behaviour of a functionally autonomous society that has maintained its existence through a sufficient number of generations so that each stage of the life span of an individual is included within the system. Such learned behaviour, when studied, has been found to be systematic, and this systematization can be referred to the uniformities in the structure and the functioning of the human beings who embody the culture. While the content of any particular culture has to be referred to a long series of historical events, many of them fortuitous from the standpoint of any given culture, the form in which this historically determined content will be expressed will—given a sufficiently long period of time—become systematic and, as such, will be comparable with the same formal aspects of other cultures.[2]

Underlying all the scholarly definitions of culture, from learned behaviour, 'intellectual development', axes and pottery, to culture as a function of civilization, there are three basic assumptions.

Firstly, modern social sciences consider culture to be purely mechanistic, and as such, essentially a product of human behaviour. That is to say, behaviour delineates culture but culture does not determine human behaviour.

Secondly, following from the mechanistic and behaviouristic outlook on culture, it is assumed that there is no place or role in culture for human free will. After B. F. Skinner, it is considered that man is totally dependent on his environment and reacts only as a reaction to his environment.

Thirdly, it is assumed that the concept of culture is virtually value-independent and outside the influence of any ethical or moral code.

In contrast, culture in Islam is not a mere sum total of actions, reactions and behaviours or a complex of art, literature and sciences or even fashions of life or modes of existence. These in fact are the end products and manifestation of culture, but not culture itself.

In Islam, culture is an attitude of mind, a mental outlook, a world-view. Islamic culture is a manifestation of being in a state of Islam. It incorporates a tradition as well as a historical experience. When one drops this tradition, or loses this experience, or blocks out both with the Coca-Cola mentality or when this tradition and experience are only in the unconscious, the opportunity to build a solid society in the contemporary setting is wasted.

To insist on the preservation of a culture, to practise a tradition does not mean the worship of the past; on the contrary, a tradition is alive only when it is flowering. As far as Muslims are concerned, this requires developing the mental outlook that is the culture of Islam.

What is the basis of this outlook? It lies in the fundamental concepts and beliefs which are the foundation of Islam—namely, the belief in the unity of God, the Prophethood of Mohammad, the life after death, the dignity of man, and the unity of mankind. The central belief here is the belief in God; once this is accepted the rest fall in place. But belief in God presupposes an understanding of His true nature and attributes. Without this understanding it is not possible to have a clear concept and a meaningful belief in God.

The Qur'an corrects and clarifies various human misconceptions about the true nature of the Deity, the *Ilah*, and presents a complete and composite view. He is One and without any associate or kin, Divine in His own right, Absolute and Almighty, All-knowing and All-seeing, Omnipresent and not confined to any object, place or precinct, Kind and Just, Merciful but capable of catching and punishing the guilty. It is only He who can confer or withdraw any favour. If He gives none can stop, if He stops none can give. The significance and relevance of *Iman billah* (belief in God) can very well be seen in the context of human thought and endeavour.

In believing in God, the whole area of human thought and vision becomes practically co-extensive with the cosmos. Instead of being narrow and limited, the outlook and approach becomes truly universal. There is a unique combination of a feeling of dignity as well as a genuine humility. Once a person comes to understand the true nature and relationship between the Creator and the created, he is at peace with himself. He is optimistic, patient, brave, content and satisfied. But the greatest advantage of *Iman billah* lies in the social and cultural sphere. It confers an exceptional and sublime purity of motivation to human actions, creates a genuine discipline and desire to order one's life according to the dictates of Allah, and generates the best in human feelings and efforts. Nothing like this can be achieved or generated by applying any or all the other disciplines of persuasion or coercion.[3]

This belief in God gives rise to a particular attitude of mind which moulds the social, intellectual and spiritual consciousness of a Muslim. This is the highest place in the sphere of significant human development; and it contains the elements of certain varieties of open systems of enquiry. It pre-supposes freedom from all prejudices and presumptions and recommends a critical and rational approach to the study of life as well as observation of nature and critical appreciation of its laws. If the Muslim's way of life merely alienated him from reason, without providing him with a framework for its proper use, then it would be nothing more than a refuge for poetical self-indulgence. But this is not so. The mental state generated by the religion of Islam represents a cosmology. It represents a balanced approach to life; it possesses a dynamic sense of moderation. It seeks neither to dominate nature, nor to be dominated by it; it seeks neither to reject technology nor to be enslaved by it; it seeks neither to be overtly materialistic nor to be utterly abstinent. Here, there is harmony between the material and the spiritual. The real is reconciled with the ideal. There is a practical synthesis between hedonism and asceticism.[4]

The Muslim lives in a cosmos. In his mind the fragments of social and spiritual experience fit together into a perfect picture, like the bits of glass in a kaleidoscope. To him, the universe is a totality. He sees man in his manifold relations, surrounded by a kindred universe in which everyone fills his alloted place. He knows what is immutable and what is only transient. For him values and norms do not admit change: the values of the future will, for him, be exactly the same as values of today and yesterday. The symbol of Islam is not a flowing, meandering river; it is the stable cube of the K'aaba in Mecca.

The cosmos of the individual Muslim is no different to the cosmos of the *Ummah* as a whole. But, Islam attempts to harmonize the demands of individualism with that of collectivism. The relation between the individual and the society is such that neither the personality of the individual suffers any domination or corrosion nor is the individual allowed to exceed his bounds to such extents as to become harmful to the community as a whole. In reality there is no conflict. The quest of the individual's life is the same as that of the life of the whole *Ummah*, the execution and enforcement of the laws of Islam and the acquisition of the pleasure of Allah.[5]

Within this framework, the individual and the community seek the same ideals. Namely:

1. On the political level, the ideal includes a healthy society, guided towards well-articulated and well-defined goals, and guided by clearly outlined laws and institutions. A society governed by a universally applicable and comprehensive Divine Law. Here democracy plays its full role; but the people are ultimately subject to the Divine Law. There is popular participation in running of the society. Liberty of the individual is respected and upheld. However, democracy is not allowed to go unchecked; it is always subject to the law of the Qur'an.

2. On the social plane, the ideal is to develop a pattern of life and social order wherein there is no recognition of the distinctions of caste, colour, financial status or creed. Recognition is given to moral

virtues, sublime character and lofty personality. Furthermore, in this social setting there is to be no domination or enslavement: of men by men, of men by machine, of men by nature, of nature by men.

3. On the intellectual plane, the ideal is to cherish the love of knowledge and learning. (That is knowledge as defined by the epistemology of Islam—see Chapter 2.) In Islam, 'the ink of the scholar' is considered more sacred even than 'the blood of the martyr'. But the scholars of Islam cannot be compared with the computerized intellectuals of the Occident. These translate knowledge first into information, then into 'bits' of information which they assemble for particular tasks. The Muslim scholars combine knowledge with wisdom. Wisdom is 'the capacity of judging rightly in matters relating to life and conduct; soundness of judgement in the choice of means and ends; sometimes, less strictly, sound sense, especially in practical affairs'.[6] As such wisdom implies the ability to discriminate qualitatively among different manifestations of thought and among varieties of choices relevant to a particular situation. The intellectual ideal of the Muslim society is to be in a position to make choices of priority, to implement technological and social change with wisdom.

4. And finally, there is the moral ideal. The belief in justice and righteousness; universal moral values, the pivot of the ethics of Islam. Abdul A'la Maududi explains this ideal in terms of the relationship between God, universe and man:

> This concept of the universe and of man's place therein determines the real and ultimate goal which should be the object of all the endeavours of mankind and which may be termed briefly as 'seeking the pleasure of God'. This is the standard by which a particular mode of conduct is judged and classified as good or bad. This standard of judgement provides the nucleus around which the whole moral conduct should revolve. Man is not left like a ship without moorings, being tossed about by the blows of wind and tides. This dispensation places a central object before mankind and lays down values and norms for all moral actions. It provides us with a stable and flawless set of values which remains unaltered under all circumstances. . .
>
> Islamic culture is thus a unique combination of motives, thought and social action. It brings out and mobilises the innate goodness in man and turns that goodness into a dynamic

social force.[7]

Any development in Muslim societies which tends to check this social force, or takes these societies away from the ideals of Islamic culture is un-Islamic, and unacceptable. Yet this is what has happened. The Muslim *Ummah* has accepted and adopted development patterns based on occidental value systems. The outcome of this distortion of Muslim values is well reviewed in the prospectus of the Muslim Institute:

1. Muslims have for about 200 years suffered a period of continuous and rapid decline in all fields of human endeavour — economic, social, political and intellectual — and have been surpassed by a rival and mostly hostile civilization of the West.

2. After over 100 years of acquiring Western education, adjusting and adapting to Western political, cultural and intellectual domination, undergoing the transition to nationalist/secular identities and capitalist/democratic philosophies, and the headlong pursuit of European-style modernization, there is as yet no end in sight to the Muslims' relative and absolute decline.

3. Conversely, after about 200 years of sustained growth, development and world domination, the Western civilization (including the communist experiment) has predictably failed to provide mankind with a viable framework for social harmony, moral and spiritual fulfilment and satisfaction and international peace; Western civilization has in fact created more problems of greater complexity for mankind than those it may have solved.

4. The Western civilization's greatest achievement — the production of goods and services on an unprecedented scale — has, because of the nature of social relationships in both capitalism and communism, destroyed the moral fabric of the human personality and society and has led to mortal conflicts at social, political, economic and international levels.

5. The social relationships of Islam, on the other hand, would allow for even greater material well-being in a harmonious social order which is also free from conflicts between men, groups of men, factors of production or nations.

6. The Muslims' quest for 'modernization' and 'progress' through the Westernization of Muslim individuals and Muslim societies was, therefore, bound to fail and has done so at great cost to Muslim culture and the economic, social and political fabric of Muslim societies.

7. The damage to Muslim societies is so extensive that it may not be possible, or even desirable, to REPAIR and RESTORE their existing social orders; the only viable alternative is to CONCEIVE and CREATE social, economic and political systems which are fundamentally different from those now prevailing in Muslim societies throughout the world.[8]

Social Structures

One aspect of this damage manifests itself in the social structures that we find in many Muslim countries. These social structures result from the gulf between the traditional culture and the imported occidental variety as well as the occidentalized patterns of development. Following Kalim Siddiqui[9] part way, we find four categories of 'social class' in Muslim societies, namely: the capitalists, the landlords, the bureaucrats and the traditional scholars who go under various dubbings including 'the mullahs' and 'the shaiks'. The first three are part of the occidentalized sector, while the traditional scholars stand for all that *they* consider to be within the circumference of Islam.

The occidentalized sector was, of course, nursed and nourished by the colonial administration which made explicit forms of social stratification on racial distinction, the possession of certain technical and administrative skills, and the appropriation of central political power. The dominated Muslims could do little about the racial differentiation, but many saw acquisition of occidental skills, particularly European education, as a prerequisite to challenging the political hegemony of the colonial rulers.

All this is not to say that traditional Muslim societies were 'classless'. Many traditional communities indeed exhibit quite distinct social stratification mainly introduced by a corrupt system of monarchies. These systems supported numerous dignitaries and state functionaries who were accorded a high social status.[10] Needless to say, this type of stratification has no place in Islam where the Caliph and the beggar are on the same social footing, entitled to the same social status. However, these ideals were not always translated into reality. Colonial administration magnified what was worst in traditional societies and introduced stratification where there was none. The whole process was taken a step further in the mid-twentieth century. Urbanization, industrialization and the integration of the local economies into a world market all introduced new social groupings that were outside the traditional system of social control. In many rural areas traditional societies were forced to transform from largely subsistence production to market or export production—a process that allowed the development of considerable differences in wealth and status.

The modern day 'capitalists' and landlords in the Muslim world by no means have a unified character. We can distinguish three broad categories. The first, and historically the oldest, originated with the aristocracy, the merchants and the businessmen who had wealth and status before as well as during the colonial administration. The second group consists of a more aspirant stratum educated and trained in occidental skills but not of established status. Some acquired considerable wealth from their professional activities and later moved into industry. Others moved into politics both as a means to reinforce their wealth and to buttress their status. The third group consists of the opportunists who exploited the occasion of independence.

The bureaucrat is a descendant of the pre-colonial monarchies and largely a product of the colonial administration. This administration set out to produce a 'go-between' between the rulers and the ruled. What Macaulay wrote in 1835 in his Minute of [Indian] Education, was also echoed by other colonial powers: 'We must at present do our best to form a class who may be interpreters between us and the millions whom we govern; a class of persons, Indians in blood and colour, but English in taste, in opinion, in morals, and in intellect.'[11]

This neatly sums up not just the bureaucrats, but also a large segment of the occidentalized elite. The bureaucrats place much emphasis on the hierarchical character of this inherited apparatus; this distinguishes the 'senior service' and junior employees. Each category is of a different 'grade', has specified duties, different pay-scales and a distinct status in work situation.

Together, the capitalists, the landlords and the bureaucrats form the ruling elite of the Muslim world. There are three elements which bind them together: (1) possession of the means of 'legitimized' violence and coercion—namely political power, armies; (2) possession of a sense of inherited legitimacy—that is, a symbolic right to take over the colonial administration; and (3) possession of the manipulative skills of the government. The elite group uses these elements to reinforce their position. The three groups provide mutual support: capitalists finance political parties; bureaucrats back leaders who do not challenge their position; government contracts are only supplied to supporters of the regime. Wealth acquired from political office is used to acquire land, property, or industries. Those who raise their voice against this system are mercilessly repressed.

In general, political regimes leaning towards various shades of Marxist-type occidentalism have been more ruthless. The brutality of Marxists in Syria and Iraq, the People's Democratic Republic of South Yemen, and Somalia is well known. In contrast, conservative governments, with the notable exceptions of Iran, have been more humane and more concerned for the preservation of tradition and culture.

Nevertheless, wherever there is an occidentalized elite in power there is a strong oppostion from the traditional sector. Kalim Siddiqui places this opposition in a historical setting:

> The dominance of these elites in their countries was not as complete and unchallenged as they and their European mentors had supposed. European rule had left the traditional values of these societies intact; in some cases, prolonged isolation of deeply held values from government induced socio-political and economic change and may have made traditions stronger and less receptive to change. The thin layer of modernity laid over basically traditional societies through a small Westernized elite produced an illusion of Europeanization.[12]

In their response to the political and economic domination, first by the occidentals, and then by the occidentalized elite, the traditional scholars often went to the extreme. They completely shunned all forms of occidentalism and dedicated themselves to the maintaining of Islamic culture uncontaminated and free from alien influence. Perhaps in their zeal they went too far; to the extent of cutting themselves off from contemporary reality.

> But the fact also remains—and the 'modern' Muslim would do well to recognise it—that had it not been for the almost fanatical zeal and selfless dedication of the 'traditional' Muslim, perhaps the crisis of identity of the Westernized Muslim today would have been even more acute than it is. The 'traditional' Muslim at least has kept the candle burning through the truly dark ages of Muslim history. That we can still see a glimmer of light at the seemingly endless tunnel of our dark ages is only and perhaps solely because the candle in the mosque has never been allowed to go out altogether.[13]

There are three main reasons why the traditional sector has always opposed the occidentalized elites:

1. Traditional scholars are inherently conservative. They do not accept change for the sake of change and have to be convinced of

the necessity as well as value of particular change.

2. They hold the occidentalized elite responsible for the decline and decay of Muslim societies. (To be fair, the process of decay had well set in before some Muslims turned towards occidentalization). An often quoted accusation concerns the occidentalized elites' concern with the imposition of the colonial language as the language of the administration and government. This insistence is based on the presumption that linguistic uniformity fosters development. But it also ignores one profound wellspring of change: a society's pride in its language, and its desire for cultural esteem. The traditional scholars regard the destruction of language as cultural genocide.

3. They are eager and anxious to preserve the cultural identity of Muslims and safeguard it from all erosion and subversion. Indeed, their resistance seems to increase proportionally as the subversive effects of the occidental imports become more and more realized.

All these reasons are understandable, and worth respecting. Indeed, the occidentalized Muslims are beginning to appreciate the position of the traditional scholars. Perhaps this is the result of their disillusionment with the occidental civilization and their own flirtation with it. Nevertheless, they are becoming more aware of the need for cultural preservation and are increasingly motivated to commit their own human, material and intellectual resources to the collective goals of the *Ummah.* [14]

On the other hand, the traditional scholars are acquiring more and more modern skills and techniques. It is common nowadays to come across traditional scholars who have also gone through the occidentalized systems of education. As such their understanding of the occidentalized Muslims is increasing.

All this will inevitably bring the two sectors together. Indeed this is already happening to some extent. The two sides are losing the contempt in which they have held each other and recognize that each is a peculiar product of the era of decline of the Muslim *Ummah.* There is an awareness for the need to combine the resources—intellectual, moral and spiritual—of both sides for devising acceptable strategies for science policy and economic development.

Ethnic Diversity

The *Ummah* we noted in Chapter 1 consists of Muslims from all types of ethnic backgrounds. Islam recognizes ethnic differences as something of the greatness of God rather than the greatness or superiority of one colour, or language, or race, or nation over the other. The Qu'ran states:

And among His signs are the creation of the heavens and the earth, and the variations in your languages and your colours: truly in that are signs for those who know.[15]

And again:

Mankind! We created you from a single pair of male and female and made you into nations and tribes that you may know each other (not that you may despise each other). Indeed the most honourable of you in the sight of God is the most righteous of you.[16]

While the standard of human worth and nobility is thus based on God-consciousness and righteousness irrespective of one's ethnic or tribal origins and affiliations, Islam does not ride rough-shod over these distinctive ethnic characteristics. These are meant to promote recognition instead of exclusiveness and arrogance implicit in the concept *'asabiyyah* (blind group solidarity/racialism/nationalism) which is condemned in Islam. There is a saying of the Prophet to the effect that whoever summons, fights or dies for *'asabiyyah* is not a Muslim. There is thus the paramount emphasis in Islam on unity based on nobility and equality, and brotherhood within ethnic diversity. Many ethnic groups have traits that are undoubtedly secondary and even trivial. But there are many which are more central. Each ethnic group may have characteristics unique to itself—language, dialect, culinary habits and gastronomic tastes, traditions of hospitality and courtesy and so on which (so long as they are not repugnant to the values and principles of Islam) all contribute to the richness and beauty of the human social tapestry, a richness and a beauty which is apparent in all of nature's organic species and which cannot be expressed in a single form without suffering debasement or loss.

In the face of rapid development trends some of these distinctive relationships and characteristics are smothered or erased—to be lost irretrievably. When Sindhis oppress Baluchis (or vice versa), when Kurds are shuttled like cattle between Iraq and Iran, when the nomad is forced into settlements in the Middle East, the general quality of Muslim life suffers diminution. Irreplaceable ethnic patterns are wiped out. The victims are reduced to socially impotent status and thus prevented from flowering culturally. The aggressors demean themselves in the very process of demeaning others and thereby lower their own human quality.

Many ethnic minorities in the Muslim countries have suffered heavily from the occidentalized patterns of development that are prevalent today. The Bedouins in Algeria, the Pathans in Pakistan and the Kurds in Iran and Iraq are but a few examples. The vast differences in political influence which any one ethnic group, or more accurately its leaders, can have on the political process for the group's own advantage, are apparent to any observer of the Muslim world. However, it should be emphasized that whatever picture of balance of dominance disability between ethnic groups is envisaged, it only has a temporary character, for any balance may be upset by the outcome of inter-group conflict. The civil war in Pakistan, and the reduced fortunes of Bengalis is a grue-some reminder of this. But politics apart, there are four variables of the present occidentalized development patterns that help determine the differential ranking between Muslim ethnic groups:

1. The size of population and the extent to which natural resources are located in areas occupied by a particular group may lead to considerable disparities between groups, particularly where political and constitutional arrangements bring this factor into prominence.[17]

2. The acquisition of modern techniques and social skills within each group may be vastly different. Here the degree of adaptability to and the use of advanced methods in agriculture, industry and commerce by particular groups is important. Of further importance is the adaptability and receptivity of the group to occidentalized forms of social control.

3. A relationship of dominance and subservience often arises by (conscious or unconscious) use of pre-colonial patterns of conquest or from the unequal division of labour where a particular group may enjoy a monopoly of specialization in a valued skill or trade.

4. Under the colonial order, many different patterns of recruitment in certain occupations were introduced. After independence these patterns were uncritically accepted by the occidentalized bureaucrats and reinforced further. Thus certain ethnic types are still identified as only suitable for a certain type of work. The Pathans were considered good soldiers, the Bedouins good labourers and nomads elsewhere good for nothing, and so on.

There are, however, two important reservations that must be borne in mind regarding the above discussion. Firstly, our description of inter-ethnic conflict in Muslim societies is somewhat oversimplified in that it assumes that groups act collectively, as if with one voice. In practice, this is far from the truth. It is usually the occidentalized leaders of the particular groups who articulate 'grievances', and whip up mass feelings, which eventually lead to conflict. Between the leaders and the led there is a whole range of actors who make their own interpretations and make their own decisions. Secondly, the role of occidental powers in creating ethnic differentiations within the Muslim countries must be fully realized. These include the following: the key occidental commercial interests, the de-colonizing powers, and those powers which have key strategic or economic interests in the country concerned.

From the cultural and ethnic aspect, development should be seen as a goal-orientated, multi-dimensional movement capable of unleashing the evolutionary potential of Muslim societies. Such a movement is essentially a cultural process with three components: the socio-political organization, the ethnic diversity and internal harmony, and the intellectual-ideological systems of a society. For the Muslim societies these three cultural components of development must be based on the ideals of an Islamic society.

NOTES

1. L. A. White, 'The Concept of Culture', in M. F. Ashley (ed.), *Culture in the Evolution of Man* (Oxford University Press, 1962), p. 38.
2. Margaret Mead and Rhoda Metroux (eds.), *The Study of Culture: At a Distance* (University of Chicago Press, 1949), p.22.
3. M. H. Faruqi, 'On the Basis of Culture in Islam', *The Muslim* 8 (4) (Jan./Feb. 1971), p. 81.
4. See M. Marmaduke Pickthall, *The Cultural Side of Islam* (Ashraf, Lahore, 1961), ch. 1.
5. Cf. M.Z.Siddiqui, 'Islamic Culture—What Do We Mean by It?', in *Islamic Culture. A Few Angles* (Ummah, Karachi, undated), p.45.
6. C. T. Onions (ed.), *The Shorter Oxford Dictionary in Historical Principles*, 3rd edition (Clarendon Press, Oxford, 1968), p. 2436.
7. Abu Ala Maududi, *Islamic Way of Life*, (Islamic Publications, Lahore, 1965), p. 4041.
8. The Muslim Institute, *Draft Prospectus* (Open Press, Slough, 1974), pp. 3-4.
9. *Conflict Crisis and War in Pakistan* (Macmillan, 1974).
10. C. S. Whittaker argues, from the Marxist point of view, that the present traditional elite, with its authority firmly entrenched in traditional social control has, in the particular case of Northern Nigeria, successfully adapted to the new focus of political power introduced by the colonial and post-colonial authorities. While this may be true of Northern Nigeria, in the rest of the Muslim world, traditional scholars have lost out heavily to the occidentalized elite. See his *The Politics of Tradition* (Princeton University Press, 1970).
11. Michael Edwards, *Raj* (Pan Books, London, 1969), p. 151.

12. Kalim Siddiqui, *Functions of International Conflict. A Socio-Economic Study of Pakistan* (Royal Book Co., Karachi, 1975), p. 7.
13. The Muslim Institute, *Draft Prospectus*, p. 14.
14. Ibid., pp.13-15; also Kalim Siddiqui, *Towards a New Destiny* (Open Press for the Muslim Institute, Slough, 1974), ch. 4.
15. The Qur'an 30 : 22.
16. Ibid., 49 : 13.
17. A. M. Khusro, 'Economic Laws of Majority Minority Relations'. Paper presented at Muslim Institute Seminar, 21 August 1976 (to be published). Professor Khusro shows that even a small bias on the part of the majority results in heavy loss of opportunity and eventually talent on the part of the minority.

5 THE SOCIAL SIDE OF DEVELOPMENT

Urbanization is one of the main indications of modernization. Daniel Lerner has used urbanization as 'an index of modernity'. He constructed a matrix containing data on urbanization along with literacy, voting and media participation and used it to calculate indices of public participation in the four 'sectors' by expressing the data as the proportion of total population possessing each attribute. The higher the level of urbanization, the more 'modern' the society.[1]

That was in 1958. Today, urbanization has taken on crisis proportions: substantial towns doubling and trebling their size in a decade have become an accepted pattern. Algiers, for example, grew from 310,000 inhabitants in 1950 to 900,000 in 1960. Accra and Lagos record equally massive increases, while Karachi, Djakarta and Tehran doubled in that period. How such vast aggregates of humanity can be accommodated is almost beyond comprehension. However, one thing is certain: urbanization is not leading the Muslim world towards the promised utopia of 'modernization' but towards a collapse of the life support systems and the doom of many Muslim cities as functioning systems.

Why is urbanization increasing at such a rate? First, compared with the experience of Western Europe, not only are all the development processes in the developing countries being greatly accelerated, but also they are spatially concentrated. The conventional development strategies demand emphasis on industrialization and mechanization. However, attempts to enlarge the industrial sectors are often choked by the poor level of infrastructure available. The basic requirements of industry — electricity, water, road, rail and post facilities, banking and insurance — can only be found in large cities. Often the only city which can provide the back-up facilities for industry is the capital. Once industry has been established in a city, its presence enhances the drawing power of that town. A cycle is now set in motion: the more industries that congregate in a city, the cheaper it becomes to supply their infrastructural needs. Dakar, for example, supports eighty per cent of Sénégal's workers in manufacturing industry and seventy per cent of the country's commercial workers.

The increase in urban population has, of course, the other side of the equation: a decrease, or a 'push' from the rural area.

The entire process of increase in urban population and decrease in rural population is assisted by the following policies pursued by many Muslim countries:

1. Industrial licensing, which encourages the existing industrialized areas to grow even faster (rather than encouraging more decentralized industrialization).

2. Transportation policies which favour the movements of people and goods towards the industrial complexes.

3. An almost total lack of urban, as well as rural, planning. In most Muslim countries, the urban forms, city governments and urban and rural planning schemes are archaic, occidentalized and have little relevance to the problems of the society.

These policies not only accelerate the rate of increase or urbanization, but also usher in two related problems: unemployment and poverty. At present 'visible' unemployment in the Muslim countries ranges between 10-15 per cent. But when estimates include low productivity, underemployment — often called hidden or disguised unemployment — the problem's proportion can be seen to be staggering. In the rural areas the fixed notions of 'unemployment', with all its implications of set hours, formal wage and registers of unemployment benefits cease to have any meaning.

The mistake by economists and other social scientists has been to conclude that because an individual in an urban area of a developing country is seeking a wage job but unable to find it, he is unemployed. In an industrial country, this is valid for the reason that casual wage employment and self-employment opportunities are severely limited, and those that exist would provide an income far less than that of the vast majority of working population. This is not true in developing countries. In most the possibilities for casual and self-employment are extensive (indeed, such employment is the basis of the economy), and while the income derived from these pursuits in many cases is less than that for wage jobs, it approximates to that earned by peasants on the land, where most of the population is. If all those without wage jobs in African towns were to be considered unemployed, then we are not talking about rates of 20-25 per cent, but of 70-90 per cent. But, of course, it is absurd to define the employment norm in terms of the conditions enjoyed by a tiny majority of the labour force — this has not, however, deterred social scientists from doing so — better the irrelevant concept one can fit easily into *a priori* models than the relevant one that is not subject to partial differentiation.[2]

The concept relevant to the rural areas is, of course, poverty. Poverty is detested by Islam. Yet with the exception of a few oil-rich nations, large segments of population in most Muslim countries suffer from poverty. Poverty is not just misery, but a savage kind of hell. It has two major components: hunger and impotence. About hunger, the Indian author Kamala Markandaya wrote: 'Hunger is a curious thing: at first it is with you all the time, working and sleeping and in your dreams, and your belly cries out insistently and there is a gnawing and a pain as if your very vitals were being devoured, and you must stop it at any cost ...Then the pain is no longer sharp but dull and this too is with you always.'[3] The other component of poverty is the demoralizing sense of personal and societal impotence that it generates. One cannot understand poverty in terms of statistics or just by gazing upon it. Only experience can teach one its real meaning.

As a function of need, poverty has come to mean different things to different people. In 'primitive' time, poverty was unknown. Even today Australian tribal communities living the life of Stone Age people experience no poverty. It is only when these tribes come in direct contact with the occidental societies that want and poverty are thrust upon them.[4]

In Western Europe, poverty as we now understand it became widespread only after the Industrial Revolution. The development of industrial production forced on the rural masses a completely new lifestyle and placed them in a drastically changed physical and social environment. The towns and cities of nineteenth century Western Europe are well known for their overcrowded, insanitary conditions, high mortality rate and the subsequent break-up of rural culture and social life.

The changing context of life led to a kind of accumulating debt — or poverty — in one aspect of existence after another. Society exchanged poverty in one sphere for poverty in others which seemed less vital. Problems of transport, urban sanitation, entertainment, education and social activity became increasingly pressing. It is to the development of problems such as these that we owe much of the modern urgency of consumption. New needs spring from the changing lifestyles and environment in two ways: the old methods of satisfying many human needs were destroyed or rendered obsolete and new problems demanding new solutions arose out of the new patterns of living. Society has relied heavily on the economic system to right this situation. Man has had to be encapsulated increasingly in his own creations to make his industrial life-style workable.[5]

What in fact happened amounted to a simple substitution of rural poverty for urban poverty. Rural poverty is generated by the absence of material to meet the lowest need of an individual: the need for food and sustenance.[6] Urban poverty is created with a dehumanized environment of a kind quite different from the already appalling poverty of the countryside.

In many Muslim countries, particularly those of Africa, both types of poverty are rampant. Can we measure this poverty on a quantitative scale? Any attempt at measurement of poverty meets lack of accurate statistics, difficulties in attempting to include relevant items, accounting for cost of goods and production and so on. One way is to use per capita output of a country as an indicator. Here the differences are so large between many Muslim countries and the countries of the Occident that virtually any measuring tool would bring out the disparity. Table 1 groups Muslim countries according to their annual per capita output in US dollars at the countries' exchange rate. However, foreign exchange rates reflect the price of internationally traded goods and, as such, over-emphasize the gap between those with 'too much' and those with 'too little'. As an example, $20,000 could purchase a one-bedroom flat in London's suburbia. In one of the African or Asian Muslim countries, the same amount can buy a decent house. And again: 10 cents can buy a filling meal in Pakistan, Nigeria, Indonesia. What can one buy for a dollar in Europe or America?

One of the principal causes of urbanization is the considerable differential between economic opportunity in the city and country. The power of trade unions in the large conurbations is becoming evident in pressure on urban wage levels. Government employees in the public sector are usually the best organized and their wage increases act as a base for a wide range of economic activity as well as a pattern of urban wage increases. The rural economy is based substantially on subsistence agriculture and the gap between rural and urban income continues to grow.[7]

The inequality of opportunities between town and country also

Table 1:   Annual per capita output of some Muslim countries.

Group A: Annual per capita output of $0 - $100

Malawi	Somalia
Mali	Upper Volta
Mozambique	Afghanistan
Niger	Yemen
Sénégal	Bangladesh

Group B: Annual per capita output of $101 - $300

Algeria	Indonesia
Nigeria	Iran
Liberia	Iraq
Morocco	Jordan
Sudan	Malaysia
Tunisia	Pakistan
Egypt	Syria

Group C. Annual per capita output of $301 - $ 600

Turkey

Group D. Annual per capita output of $1,600 - $3,000

Bahrain	Denmark*
Libya	France*
Oman	Norway*
Qatar	Switzerland*
	United Kingdom*
	USSR*
	West Germany*

Group E: Annual per capita output of above $3,000

Kuwait	Sweden*
Saudi Arabia	USA*
United Arab Emirates	

* For comparison

---

manifests itself in housing, education and medical care. Although there is an all-round shortage of housing (and not only in developing countries), the chances of obtaining shelter are much higher in the urban areas. In rural areas the housing problem can be met by largely traditional methods. However, the complete reliance on modern construction methods which depend on materials not readily available augments the problem.[8] In the domain of education, a child from a Pakistani city is eight times more likely to go to university than a village child. Thus, luck in having an urban birthplace determines a child's chances in life. Apart from equity, such a system proves to be quite inefficient as many university places are being used to train less able urban children, rather than more able rural children.

The same inequality appears in medical care. City dwellers have three to four times the chances of obtaining medical attention. This is because governments of Muslim countries place emphasis on large, central hospitals and urban medical services. The need, however, is for small, district hospitals, rural clinics, paramedical staff and environmental health. Four-fifths of doctors in the Muslim countries operate in towns. Rural service is considered no better than exile to Siberia.

We feel that there is a great need for Muslim countries to change their outlook towards medicine. The present immobile institutional framework and heavily technologically orientated medicine has an almost dehumanizing emphasis. This system of medicine concentrates on cure rather than prevention. Islam, on the other hand, places greater emphasis upon preventive medicine rather than on curative medicine.

> Islam stresses the permissible, but even more so it emphasizes abstinence from the prohibited. The permissible element has a deep relationship to preventive medicine: but the prohibited element exercises an even greater effect. Harmony between the soul and the body is necessary for following the edicts regarding the permissible and the permitted.[9]

As such, there is a need for much emphasis on preventive medicine and environmental health. This involves a swing towards a much wider socio-economic, de-centralized as well as traditional medicine. In particular, following S. M. Rashid in parts, there is a need for:

1. Community medical care programmes which develop large numbers of small, local hospitals, dispensaries, clinics, emergency care units, and the required administrative machinery and physical

facilities.

2. Community preventive services, for example: environmental health programmes, collection and use of medical and vital statistics, child care units, health services for schools and factories, and communicable diseases control projects, etc.

3. Health manpower training, for example: training of paramedical staff, undergraduate and post graduate training of various categories of health workers.

4. Medical and social research on rural health and environment.

The underlying issue in the simplification and decentralization of medicine concerns the devising of alternative delivery systems. In this regard the Muslim countries are fortunate in having alternative systems of medicine available in the traditional systems of medicine such as the *Unani*, which have developed not just sophisticated institutional forms, but also a healthy environmental approach to medicine. In the re-orientation of outlook on medicine, these traditional systems have a great role to play. The endeavours of the Hamdard Foundation of Pakistan are well known, as are its successes in developing traditional remedies for modern maladies. The two systems of medicine are not in conflict with each other, rather they are complimentary. It is indeed a pity that this has not been realized, outside China.

The standard approach of the occidentalized institutions to the problems of the urban poor is to assume that their needs can be met by cheaper or cruder versions of imported urban technologies. This assumption overlooks the fact that rural life and environment form an interrelated tightly woven web. In this web all the parts are closely related to all the other parts. Only when seen as a whole does it really make sense. Indeed what is important is the whole system and the set of relationships that it contains. Yet it is the constituent parts through which the system has evolved and which give it a reality. The imported technologies upset the balance between the traditional life-styles and the environment thus destroying the whole system. They also accentuate the dependence of the rural areas on the urban settlements.

The occidentalized planners have no regard for traditional technologies which they equate with 'primitiveness' and, often, 'savagery'. Their disregard results in a tragic loss of centuries of craftsmanship and knowledge. The urban technologies, because they are mass producing and labour saving, destroy the business of the traditional craftsmen: this affects not just the business of a few individuals but also terminates the long-established tradition of a craft. The death of business can also mean the death of a craft.

The opportunities available in the rural environment also affect the total pool of talent in the villages. As there are no job opportunities within the villages, nor any opportunities for taking up a particular craft, the village youth have either to migrate to the cities or take up farm labouring as a life-style. Thus the rural areas lose most of their talent; and the little talent that remains diminishes rapidly for what is talent but (largely) a function of opportunities?

The rural environment reacts to the destruction of the traditional way of life. The ability of the environment to cope with new stresses is reduced, and any interaction between social, economic and political factors, which are usually long term in nature, can create a condition for disaster; that is, they make it inevitable that at some point a breakdown will occur.

> The condition for disaster can exist for a long period of time before catastrophe strikes. Very often it is the impact of some natural phenomenon which overloads the system. It is this close association between breakdown and natural phenomena which gives rise to the tendency to describe these events as 'natural' disasters.[10]

The fact is that there is a strong link between occidentalized development strategies, dependency and underdevelopment. Famines, droughts, and other disasters are all man-made and can be traced to occidentalized development strategies. The commercialization of crops leading to monocultures or digging of wells to regularize the movements of pastoralist populations, for example, have made large contributions to the de-stabilizing of ecosystems. These activities have increased the probability of disaster by exacerbating the impact of natural phenomena on an area by weakening its ecosystem.[11]

A particular cause of famine and poverty is the unemployment situation in the developing countries. Here the difference in opportunities between the rural and urban populations can sometimes result in disaster. In the case of water shortages, for example, rural

> ...labourers and small operators who have to work on wage-farms for their livelihood find their employment opportunities shrinking because the wage-farms, in the absence of an adequate water supply, try to economise on water by concentrating cultivation on smaller areas of land. This reduction in employment comes at a time when the small operator's output from his own farm has either vanished

or has been drastically reduced. In the meantime, the shortage of food pushes up its price beyond the means of the poor. The state, even if reliable information is reaching the capital, is often reluctant to acknowledge the existence of famine conditions since this acknowledgement tarnishes the 'image' of a country. In many countries, for lack of adequate communications, news from distant regions takes a long time to reach the capital. Even when the existence of famine is conceded, there may not be sufficient surpluses available for transfer to the stricken areas. Poor transport facilities are frequently incapable of handling a sudden increase in freight traffic even if sufficient food is made available. This happened in India, during the Bihar famine, when grain shipments could not be unloaded at the ports because of inadequate berthing facilities. The fair distribution of food may be further handicapped by inefficient and corrupt administrative machinery. [12]

The problem of unemployment, however, is not just a rural problem. In many Muslim countries, and indeed in many developing countries, the unemployed constitute larger and larger segments of population. There are three major aspects of the nature of unemployment which must be realized. Firstly, it has a number of dimensions, many affecting much larger segments of the labour force than did unemployment in Europe at the peak of the Great Depression. Secondly, causes of unemployment problems are many and diverse, ranging from industrialization policies and the Green Revolution to economic mal-planning. Indeed, tracing through the various areas of unemployment leads to a multi-dimensional analysis of the whole social and economic policies pursued in the Muslim countries, and forces us to question the validity of the occidentalized development strategy. Thirdly, the human effect of increasing unemployment in terms of poverty and growing frustration is very serious. There is an urgent need for major changes in both domestic policies and outlook towards the countries of the occident.

Population Policies

The problems of urbanization and unemployment are often related to poverty via the concept of 'population explosion'. The supposed demographic outlook for the next two or three generations has received so much attention that it has now lost its power to shock. Some development scientists and futurists see the rate of increase of population not just as a cause of underdevelopment but also as a pointer towards a thermonuclear war.[13] Such stories as Harry Harrison's *Make Room!*

Make Room!, Anthony Burgess' *The Waiting Seed* and those collected by T. M. Disch in *The Ruins of Earth* and Frederick Pohl in *Nightmare Age*, highlight a whole range of possibilities and responses to an over-crowded globe, from new forms of social control to governmentally approved sexual deviations. While works such as these bring a serious problem into focus, they also distract from the real issues.

That there are dangers in the present increases in population, we cannot deny. The demographic situation in many Muslim countries portends a grim Malthusian outcome and has to be accepted despite the fact that there are many areas of the Muslim world that are still under-populated in their human carrying capacity. However, the very fact that population increases are concentrated in poverty-stricken areas suggests that poverty has some relation to population increase. Research in some developing countries — Argentina, Taiwan, South Korea amongst others — has indicated that rate of increase of popula-tion is more a function of poverty than a cause, and that birth rate declines sharply when living standards rise for the majority of the people. Once we recognize that the 'population problem' is a function of poverty, a different perspective crystallizes.

There are two ways in which the increases in population can be approached:

1. by dealing positively with its root causes; namely poverty and unemployment; and
2. by attempting to restrict, indeed stop and reverse, this increase by such means as family planning, birth control and sterilization.

On the whole, the first approach has been largely overlooked. In a few cases when it has been adopted, it has not succeeded simply because the policies pursued have been detrimental to the very objec-tives they aim to achieve. The major difficulty is that the goals of employment creation and equal distribution of wealth do not coincide with the objectives of occidentalized development strategies. As such policies for creating employment have always been overshadowed by the goals of industrialization and economic growth.

> ...looking at natural plans of the developing countries, it was obvious that unemployment was often a secondary, not a primary, objective of planning. It was generally added as an after-thought to the growth target in GNP but very poorly integrated in the frame-work of planning. Recalling my own experience with formulation of

Pakistan's five-year plans — the chapter on employment strategy was always added at the end, to round off the plans and make them look complete and respectable and was hardly an integral part of the growth strategy of policy framework. In fact, most of the development which affected the employment situation favourably such as the rural works programme and the green revolution, were planned primarily for higher output, and their employment generating potential was accidental and not planned. There were endless numbers of research teams, our own and foreign, fixing up our national accounts and ensuring that they adequately registered our rate of growth: there was not a fraction of this devoted to employment statistics.

The employment objective, in short, has been the stepchild of planning, and it has been assumed, far too readily, that high rates of growth will ensure full employment as well. But what if they don't? A sustained six per cent rate of growth in Pakistan in the 1960s led to raising unemployment, particularly in East Pakistan.[14]

Having overlooked employment from their development planning, many Muslim countries pursue policies of introducing constraints within the society to reduce and if possible halt, the rate of increase in population. The obvious answer is family planning:

The adoption of a policy to restrict the birth rate thus becomes an essential part of economic and social development and no longer a result. Such a policy must include a massive effort of persuasion and education, and means of supplying the necessary tools. There has been considerable opposition in the past in many developing countries to family planning. In part this opposition is based on religious grounds. It is still the official position of the Catholic Church. Moreover, the leaders in some developing countries thought that sheer numbers might compensate to some extent for technical and military inferiority. It is now widely accepted, however, that excessive population growth has a negative rather than a positive impact on a country's economy and that it weakens a country both militarily and politically. Consequently in many areas the resistance to family planning has subsided. It has become possible to offer large scale aid in family planning to developing countries.[15]

Alpert is by no means alone in explaining the problems of the developing world in terms of population. It is a neat and well-tried tactic to shift the whole onus of poverty on to the shoulders of the poor themselves.

To the majority of the Muslims, the whole idea of birth control sounds a little repugnant. Reghili El-Naggar has summed up this feeling:

> The whole idea is not just distasteful, but positively harmful to harmonious social and scientific progress. Even if birth control is a solution to over-population, you cannot reach a zero population growth. But you do create another imbalance: between the active and inactive members of the population.[16]

Moreover, palliative solutions such as birth control play a major part in drawing the policy makers away from the real issues. Somewhere behind the glare and publicity about famine, drought, food aid, green revolution, family planning, sterilization, loops, pills, condoms, there is the real role played by the present economic structure, occidentalized patterns and strategies of development not only in creating and maintaining poverty, unemployment and population increases, but also fighting against them.

As the failure of the conventional approaches to development becomes widely recognized, the quest for new strategies for eliminating poverty and unemployment begins. Bitter experience has taught many Muslim countries that population policies must be related to policies directly aimed at eliminating poverty and unemployment. This realization, however, means that one must 'hop out' of the suffocating enclosure of the present economic order.

It is now widely accepted that the 'international economy' is in a state of crisis. The repercussions of this crisis have, naturally, been felt more severely by the poor countries. The acute difficulties faced by many countries of the 'Southern Sphere' have led them to the desire for creating a new economic order.

> The international system seems incapable of explaining and mastering the recent course of economic events. This is no doubt because that system is ill-adapted to the global dimension of the problems, to the legitimate aspirations of the new States and to peoples' needs. The basic tenets of economics themselves need to be reviewed in the light of the new economic, social and political facts. Whilst the aspiration after collective economic security is becoming widespread, with some parties anxious to retain or even to add to what they have gained and others desirous of attaining a better standard of living, individual nations, and even more, the international

community, are still by no means able to make reliable forecasts. The future is not under control and the rational approach to socio-economic matters, to which programming and planning were tending, is being defeated by the hazards of the situation and the complexity of the factors involved in change.

Since the future can thus be glimpsed only with a considerable margin of uncertainty, despite the efforts made to clarify things scientifically, we have clearly reached a point at which political agreement among all countries—and not decision by a handful of them—is essential in making choices which concern the whole world community and this pre-supposes a change in international political structures.

The need for such a radical change is all the greater because the political context in which the present economic order was worked out after the Second World War has now substantially altered. With the successive stages of political decolonization, new nations have won their independence but still have to consolidate it in the economic, social and cultural spheres. The emergence of many States which intend to take their full share of international responsibilities represents an irreversible change, to which the older nations must grow accustomed, not merely recognizing it officially in law but altering their own patterns of behaviour accordingly. The establishment of a new international economic order must be viewed in this new context and implies a critical examination of present international power (both institutionalized and *de facto*) and its reorganization.[17]

A few attempts have been made to articulate strategies for the proposed new economic order. One noteworthy attempt is by Sartaj Aziz who bases his strategy on the assumption that the whole concept of unified international community is no longer viable and will be replaced by new and different realignment of political and economic forces. As a result there will be much closer co-operation between developing countries and in particular the problem of poverty will emerge as a priority are of co-operation. The following framework is based on these assumptions:

(*a*) The establishment of a Commodity Bank financed or underwritten by special funds from the oil-exporting countries, to protect the interests of Third World countries in respect of trade in selected commodities and also to support food reserve policies in selected countries to provide greater food security in the Third World.

(*b*) The establishment of one or more development funds by the oil-exporting countries to support, *inter alia*, programmes for larger food production — particularly in countries which have the potential for producing more food on an economical basis.

(*c*) The evolution of special arrangements, in which interested oil-exporting countries provide financial assistance to other developing countries for increasing agricultural productions and receive payments of its loans in the form of specific agricultural exports of grains, sugar, meat, vegetables or other specified commodities within a stipulated period.

(*d*) Much closer cooperation among the Third World countries to facilitate exchange of ideas, experts and appropriate technologies through various institutions.[18]

From the Muslim point of view, there is a danger in blindly following the call of the 'new economic order', of falling into the trap of moving from one area of occidentalization (say, capitalism) to another, (say, socialism). This is simply substituting one non-Islamic approach for another.

After Hussain Mullick, we suggest that the discussion, within the Muslim world, on the new economic order should be based on the following 'hypothesis':

(*i*) The present economic systems in vogue in most of the Third World Muslim countries are nothing more than miniature Western capitalistic models, promoting income inequalities and under-utilization of domestic resources. Further the adoption of these economic systems also forces these countries to follow socio-political and economic thoughts developed in the West, much of which are alien to the spirit of Islam. Worse still, the imported thought patterns smother the local intellectual capital and leaves its influence on the periphery of the industrial Western metropoles.

(*ii*) The World of Islam can forge development along its own ideological and social patterns only if our scholars and politicians come forward in a big way to replace imported institutions, systems and thought patterns by substituting them with their own.

(*iii*) A separate identity by the Islamic world could only be acquired not through 'forced' imitation but through a process of voluntarism in which the Islamic society is free to absorb from other cultures in a 'synthetic' way.[19]

We can take Hussain Mullick's 'hypothesis' to be more or less proven 'facts'. From the Muslim point of view, this is a refreshingly sane start in the present ocean of insanity. That development strategies in the Muslim world can only succeed if they are based on the principle of Islam seems so obvious that one almost hesitates before saying so.

NOTES

1. *The Passing of Traditional Society. Modernising the Middle East* (Free Press, New York, 1958), pp. 57-68.
2. J. Weeks, 'Does Employment Matter?' in R. Jolly *et al.* (eds.), *Third World Employment* (Penguin, 1973), p. 62.
3. *The Unesco Courier*, May 1975, p. 4.
4. See M.Sahlins, *Stone Age Economics* (Aldine, Chicago, 1970).
5. R. G. Wilkinson, *Poverty and Progress* (Methuen, London, 1973), p. 175.
6. The other needs are considered to be safety, love, self-respect and 'self-actualization'. See A. S. Marlow, *Motivation and Personality* (Harper and Row, New York, 1954).
7. See Michael Lipton, *Why Poor People Stay Poor. Urban Bias in World Development* (Temple Smith, London, 1976).
8. On the average, in Muslim countries, there are six persons to a household with the rate of increase in the number of households of 2.8 per cent per year. The requirement to meet this increase is only two dwellings per thousand persons. The highest figure is ten dwellings per thousand persons. and this occurs in countries with a high urbanization and immigration rate, such as Kuwait, Saudi Arabia and Bahrain. Between a quarter and a third of these dwellings will serve as replacements for older buildings. (Based on UN statistics.)
9. Hakim M. Said, 'Al-Tibb Al-Islami', *Hamdard* 19 (1-6), 32 (Jan. - June 1976).
10. S. M. Rashid, 'Health Problems', *Hamdard* 17 (7 - 12) (July - Dec. 1974).
11. Nicole Ball, 'The Myth of Natural Disaster', *Ecologist* 5, 368 - 71 (1975).
12. Radha Sinha, *Food and Poverty* (Croom Helm, London, 1976), p. 15.
13. For example Barry Commoner, *The Closing Circle* (Cape, London, 1972); P. R. and A. Ehrlich, *Population Resources, Environment. Issues in Human Ecology* (Freeman, San Francisco, 1970); Rene Dubos, *Reason Awake. Science for Man* (Columbia University Press, New York, 1971); Gordon R. Taylor, *The Doomsday Book* (Thames and Hudson, London, 1971); A. Allison (ed.), *Population Control* (Penguin, London, 1970).
14. Mahbub Ul Haq, 'Employment in the 1970's — a New Perspective, *Int. Dev. Rev.*, April 1971, p.10.
15. Alpert, *Partnership or Confrontation*, p. 34.
16. Ziauddin Sardar, 'Quietly Fights the Don', *Impact International Fortnightly* 3 (2), 9 (1972).
17. UNESCO, *Moving Towards Change* (Paris, 1976), pp. 15-16.
18. Sartaj Aziz, 'The World Food Problem', *New Internationalist* 29, 22-3 (July 1975).
19. M. A. Hussain Mullick, 'Backlog and Development in the Muslim World', *Impact International Fortnightly* 5 (2), 10 (1975).

6 AID, TRADE AND THE NEW ECONOMIC ORDER

Economic planning in the Muslim countries, as indeed anywhere else, is directed towards certain goals. Most obvious of these goals are elimination of poverty and full employment: these two acute needs of Muslim societies should be the basis for any economic policy. However, this is seldom the case.

In the conventional views of development, the basic problem is considered to be the selection of the most effective method of raising the rate of investment or capital formation. This is the prime goal of all economic policy. Most Muslim countries are capital poor, and as both industry and agriculture require capital, the rate of investment is considered to place a limit on the degree of 'technological progress' which may be achieved.[1]

The problems associated with increasing the rate of investment in poor Muslim countries are fairly obvious. As investment usually takes reserves which can be used for consumption, the burden of accumulating capital is often painful, and beyond certain limits, it is quite intolerable. Furthermore, the burden is not only in terms of discomfort but often also in terms of present and future productivity as well.

There are a number of ways of increasing the rate of investment and all have some advantages as well as many distinct disadvantages. Beyond certain fundamental limits, all currently accepted methods of increasing the rate of investment become self-defeating.

In countries with 'labour surplus' economies, like Pakistan and Indonesia, investments can be increased by absorbing labourers at low wages into the industrial sphere. This would produce industrial profits which could be reinvested. Many occidentalized economists have suggested the use of unemployed or disguised unemployed labourers in various rural projects such as construction of dams, roads, irrigation ditches etc. The advantage is that at small cost, idle resources are used. In reality, these methods are difficult to apply. Often supply of cheap labour is choked off by poor agricultural production. Labour is at the best of times, difficult to utilize. The required managerial and technical skills are seldom there. Agricultural production runs the risk of falling if the labour is withdrawn, unless, of course, there is an entire re-organization

of the agricultural sector. To draw any advantage from absorbing unemployed labour into the industrial sphere requires a tricky balancing
act which is by no means easy to perform.

The 'balancing act' is also necessary for various other methods of
increasing investment rates. Consider the method of increasing public
savings. This is the difference between revenue raised — from taxation,
operations of nationalized industry etc. — and current public expenditure. These savings can be increased only by higher taxation (including
corporation taxation on private companies) or increased profits from
nationalized industry. Increased taxation is always an unpopular move
with the poor populace. In a great many cases they are already over-
burdened by severe taxation. In the case of the well-off, it could
deplete potential source of savings, could penalize economic success
and, as is often the case, if the rich have political power, it may be
difficult to enact in the first place. Then there are problems in actually
collecting the taxes. Farmers are notoriously difficult to tax; and
where the production does not come to the market, as often it does
not, the problem is likely to be very serious indeed.[2]

Nationalized industries usually run with heavy losses — due to lack
of skill and general inefficiency of both management and labour as well
as political pressures and governmental interferences — which force
them to maintain prices at a level which allows only the costs to be
recovered. Much of this is due to inflation and union pressures.

Of course, inflation itself can be used as a means of capital formation. This happens when a developing country decides to increase its
investment *without* increasing taxes. As spending increases, prices rise,
the rupees/ringgit (Straits dollars)/lira decline in real value and the
physical consumption of the country is curtailed. Rising prices lead to
high profits and the economy slowly moves uphill. A little inflation of
this nature is considered to be a good thing. The problem is that 'little'
inflation often turns into a 'big', 'runaway' monster. And this has very
serious consequences. It distorts the patterns of capital formation, tur-
ning it into speculative channels. Inflation often results in loss of
foreign markets. And there is always a real danger of creating an
upward spiral of wages and prices.

In view of the problems of increasing the rate of investments and the
serious burden on internal resources of developing countries, alternative
external resources are sought.

Dynamics of Foreign Aid

In the circles of occidental development economists, it is generally believed that no developing country can obtain economic development without some foreign aid. The concept of foreign aid is based on the European Recovery Programme, or Marshall Plan as it generally came to be known. This was designed to help Europe recover from the after effects of the Second World War. Under the Marshall Plan participating countries were required to prepare detailed four-year and annual policy plans. The chief architect of the Marshall Plan, the United States, actively supported the formulation of the plans and provided large amounts of aid. For the United States it was a short step to extend its aid programmes to the developing countries.[3]

Foreign aid is given to the developing countries in two general forms: in the form of grants and loans and in terms of 'technical assistance'. The latter comes in the shape of 'experts' and 'consultants' to advise and guide the development efforts of the recipients.

Consultants are usually required to carry out feasibility or pre-investment studies and advise on the setting up of research institutes and laboratories and take part in surveys of local natural resources. Occasionally, they are called upon to initiate new areas of research and guide research programmes. As we have stated elsewhere, in all of these, but to differing extents, wider talents than 'expertise' of a particular nature is desirable as well as a thorough appreciation of the local cultural values and norms, and a knowledge of the available —human, intellectual, technical, informational, organizational, management — resources. In a great many cases, the inability of the consultant to overcome his *deformation professionel* which prevents him from thinking other than as a narrowly trained specialist in terms of occidental values is crucial. For example, (this is fiction, coloured by fact) a consultant may find himself advising on whether there is a good case for building a local contraceptive industry. He will usually produce a report so generalized that those financing the industry will see no reason why this contraceptive industry is more urgently required than, say, an agriculture research institute. On the other hand, he may just rubber-stamp an already existing report. It may be that a better case could have been made for or against the contraceptive industry if the work already going on in the country in question and similar work elsewhere had been taken into account; if the local attitudes towards family planning were examined thoroughly; if the local resources available to build such an industry were assessed; and if the project had been couched in more modest terms, but allowing for future expansion as required. Of course, the same verdict may have been reached, even if the best possible case had been put forward. But the fact remains that the only way foreign consultants can be of positive help to the governments of developing countries is for them to understand and appreciate the local customs and resources. Otherwise, their advice is bound to be more harmful than beneficial.

Along with 'experts' and 'consultants' come those with specialist knowledge in areas other than technology. The 'technicians' sent by the ˙Soviet Union to Egypt and Turkey to advise on various projects contained a high percentage of serving officers of the KGB and GPU. Similarly with the Chinese technicians working in Pakistan and Mozambique. The activities of the American CIA are well known in this matter.[4]

Tibor Mende has compared foreign economic aid to an artichoke.

> When in flower it is fairly attractive in colour. With time it becomes a prickly plant with merely a small part of it edible. Esteemed by specialists, it also has its enthusiasts. One of its ingredients is even believed to have a curative effect against certain maladies. But to judge its real worth, the innumerable leaves of the artichoke have to be plucked one by one. Many can be discarded as worthless. Others contain the nutritive substance responsible for its reputation. Inside, deep down, one comes upon its small heart, which properly prepared and mixed with appropriate condiments provides a tasty reward for the effort that went into the patient removal of the more or less worthless leaves which hid it.[5]

In general, foreign economic aid consists of those 'worthless leaves' which hide the edible part of the artichoke. These leaves are also poisonous and can have a destructive effect if swallowed. So with foreign aid.

Leaving aside possible political motivation and the consequential use of aid as a lever, the fact is more than half the aid given is *lent* on interest. The element of 'aid' in the loan is the difference between commercial interest rates and repayment period (nowadays around 12-20 per cent per annum with repayment over 10 years) and the 'soft' rates charged to the aid recipient (around 1-3 per cent per annum with repayment over 50 years).

Usually aid is either tied or has various strings attached.

There are two ways by which aid is usually tied: either by obliging the recipient to spend aid money on equipment and services from the donor country or by spending the aid money on a particular project. Often the aid is tied both ways, although aid tied to the purchase of equipment from the donor is comprising larger and larger proportions of the total aid supplied. Aid tied to specific projects has been responsible for large imbalances in development programmes. Small projects of vital nature have attracted little or no aid while those involving conspicuous technology have been aided generously. As such, large dams, residential universities, planetariums, atomic energy programmes have received much aid, while urgently needed administrative buildings, small rural projects and area development schemes still await assistance. Often the very notion of a 'project' is a source of confusion. The recipient will interpret it as loosely as desirable, and the donor as rigidly as possible. When is a plant no longer a project? When the heavy equipment is set up? Or when the plant is marketing its output? Many plants set up with foreign aid still wait for the necessary foreign exchange to be untied to allow them to function.

Country-tied aid does not fare any better either. There is little freedom of choice: the best tractors for your purpose and pocket may be manufactured in Britain, but you must spend your money in the United States because that is where the aid comes from. As a consequence, you spend twice as much and obtain equipment which only fulfils your needs inadequately, if at all. You may have a surplus of sterling, but the aid agreement requires you to spend dollars. Country-tied aid is responsible for a complete lack of standardization in the developing countries. This is, of course, another factor which increases costs and produces 'mixed' plants—plants with components from many different countries with different standard specifications. Operating a plant of this nature is by no means easy.

Although both forms of aid tying can weaken the effectiveness of aid, non-project aid is generally more beneficial. Recipients can use the money to buy necessary goods from the donor and use the saved foreign currency for more useful purposes. On the other hand, when aid is tied to both country and project, the end product is often very unattractive as well as unproductive. A particular piece of technology in a particular country will be produced only by one or two companies. So one is forced to buy from a virtual monopoly at any price the donor sees fit to charge.

Of course, there are other ways—ranging from faint and barely perceptible to the blunt and overt—by which strings could be attached to aid. Aid reduction, withdrawal, or altering the terms of trade so that they harm the recipient are common strategies to alter the foreign or domestic policies of the recipient. In the mid-sixties Pakistan had considerable pressure brought against her by the United States to alter her policies towards China. Similarly, Egypt was pressured by the Soviet

Union over Israel. As all aid negotiations have to be formalized in a form of contract, the recipient is forced by the very nature of the process to accept some conditions. If a donor country is to invest large sums of money in a project it expects to satisfy itself as to the viability of the project and asks for any factors which may adversely, in its opinion, affect the project. Sometimes these factors are kept at the level of domestic policies such as local laws and regulations, price policy and taxation, but it may mean moving the recipient from one system of economy to another. Often foreign policy and defence come within the factors which may affect the project. At this point there is little to differentiate between aid and imperialism. But the relationship between aid and imperialism manifests itself in other ways also:

> Having been an observer of the Third World affairs from a distance, one knows the true image they have in the eyes of their benefactors. First, you solicit for a high interest rate, credit to buy over-priced industrial or even consumer goods, and then ask for a rescheduling of debt. Rescheduling is a euphemism to tell your creditor: 'Please Sir, give me more time. I don't have funds to repay. Meanwhile, you can have a little piece of our sovereignty.' Could you then expect to play your rightful role in world affairs? At the national level too you are in no real position to undertake any basic and badly needed socio-economic and political measures.[6]

It is possible to avoid a few strings, if aid is channelled through one of the many 'middlemen'. The biggest of these middlemen is the International Bank for Reconstruction and Development (the 'World Bank'); others are the International Monetary Fund (IMF), International Finance Corporation (IFC), Asian Development Bank and African Development Bank. When aid is channelled through one of these agencies, it is difficult for donors to 'lean' politically on the recipient countries. It is also difficult to tie aid to a country or a project thus allowing the aid-receivers to shop around for best value and channel the aid to most worthy projects.

However, foreign aid loans, with or without strings, tied or untied, from whatever source, have many inherent problems for the developing countries. In particular:[7]

1. There is the problem of mounting debts which frequently culminate in moratoria and severe curtailment of domestic programmes. For many developing countries, the debt servicing payments exceed

the total assistance flow, thus raising the question: *who is aiding whom?* At present the total debt of developing countries stands at the staggering sum of $130,000 million.

2. Loans have almost always to be repaid in international hard currencies or in gold. This diverts foreign exchange earnings to debt servicing and loan repayments instead of domestic investments. In addition, there is an attendant depletion of the countries' gold reserves. When, and if, the loans are reduced to manageable proportions, there is a further diversion of funds away from investment to rebuild reserve balances.

In addition, aid sets up a whole range and variety of damaging repercussions. Bauer has attempted to list a few:[8]

1. Aid reinforces the disastrous tendency in the developing countries to politicize life according to sympathy or otherwise with the aid donor.

2. Aid reinforces the pursuit of policies which are damaging to economic progress, and often inhuman; some recipient governments restrict the activities of economically successful minorities, and conceal from people the economic consequences of their actions.

3. Aid is often used for wasteful and unnecessary projects which result in large yearly deficits. Often such projects are continued for political reasons even when it becomes obvious that they bring no benefits.

4. Aid is often linked to balance of payment deficits of the recipients. The governments of the developing countries are encouraged towards inflationary policies and run down their foreign reserves. This leads to economic insecurity, which, in turn, compels the poorer country to ask for more aid.

5. The idea of continuous foreign aid obscures the necessity for developing countries to develop themselves. Often aid pauperizes those it purports to assist.

6. Aid reinforces occidental-type models and institutions which are not altogether appropriate for the needs of the society of the developing country.

We would like to see Muslim countries approaching external donors of resources on the basis of trade. Where it should be necessary to 'go abroad' we think that it would be preferable to go to similar countries rather than to developed countries. The trade involved, even though it may be at a relatively unexciting level, should prove a welcome stimulus to all. Where it becomes necessary to seek 'aid' then we consider one of two alternatives to be essential:

1. Aid donor countries or consortia should give aid according to a 'no strings attached' understanding. It may be advantageous to channel aid through one of the 'middlemen'.
2. Funds should be provided by one source for the receiver to purchase the stores in the open market.

When considering the provision of world resources to Muslim countries, a number of factors must be borne in mind. Firstly, it must be appreciated that many 'developed' countries have 'developing areas' which must absorb much of their resources. Indeed it is not unusual for Britain and Italy, for example, to seek aid from the EEC Development Fund for their developing areas, such as Calabria, Sicily, Wales, Scotland and Northern Ireland. In such circumstances there is only a certain amount which developed countries can give or can be expected to give to developing ones: true, few have yet risen even to the level of allocating the agreed 1 per cent of GNP for foreign aid. Secondly, most developing countries have some of their industries producing similar manufactures to the developing world. Some of these are in decline, some are thriving. Governments, often under great pressure from trade unions, will be anxious to safeguard employment at least during transitional phases or during the final decline of a moribund industry, and so a certain reluctance to import the usually cheaper competing products of the developing countries will be met. It is for this reason, which can be broadly termed the Comparative Costs argument, that Muslim countries, while they are still in the process of establishing differentiated commercial bases of their own economies, should trade largely among themselves and sort out 'who does what best?' Thirdly, in an area of decline in the amount of available resources, countries with favourable access to them will be anxious to preserve their advantage. It may well be that developing countries, simply because they are by and large primary producers, will find that what was an economic embarrassment in the past has become a valuable asset. The oil embargo, and the attendant energy crisis in the Occident has highlighted this increasing interdependence between developed and developing.

Aid through Trade

We would, however, strongly advise Muslim countries against reliance on foreign economic aid. We would give much preference to a policy of 'aid through trade'. If applied to all primary products, aid through trade would have a tremendous effect on increasing the economic self-reliance of the developing countries. The Occident, however, does not wish to give fair opportunities for trade to the developing world, particularly to the less-developed countries. For over three decades the Occident has pursued a policy of protection against the manufactured goods and even against the agricultural products of the developing countries. As a result the share of the developing countries in world trade has been declining continuously, from 30 per cent in 1950 to 21.3 per cent in 1960 and 17.3 per cent in 1970.

In spite of their official policy of refusal to grant 'aid through trade' to the countries of the developing world, some countries of the Occident are providing such aid to certain developing countries by giving them preferential treatment. Great Britain, for example, under the Commonwealth Sugar Agreement, has set fixed quotas for sugar she imports from the poor countries of the Commonwealth. Some socialist countries have also negotiated long-term trade agreements of this nature with some developing countries. However, sales to the socialist countries do not bring convertible currency. Usually the proceeds have to be used for purchases from the same country. Furthermore, some socialist countries have developed a habit of importing primary products from the developing countries in excess of domestic requirement and selling the surplus on the world market. The classic example here is that of Egyptian cotton sold to the Soviet Union after the 1956 Suez crisis. The Soviet Union resold part of the cotton on the world market and pre-empted traditional markets of Egypt in Western Europe.[9]

One aspect of self-reliance is that trade must be redirected from the markets of the developed world to trade within the Muslim world. Where can one go to meet the demands of one's community? To another community which can not only supply the demand but is at the same level. Self-reliance and trade go hand in hand: the Muslim world must set its goals, and achieve them through its own efforts.

At present the economies of the Muslim countries are so dependent on the Occident for finance and consumption that all efforts for self-reliance and growth are swamped. Horizontal trade between countries of the Muslim world has been made virtually impossible. There are a number of reasons for this:

1. There is a lack of common investment policies amongst the Muslim countries.

2. Capital and equipment are only available in the Occident. However, Muslim members of OPEC may now provide some of the necessary capital.

3. There is a historical tie between the traders of the Muslim countries and the Occident. Often these ties are a result of the colonial legacy.

4. There is an inferiority complex in the Muslim world *vis-à-vis* the Occident. A product manufactured in the West commands a higher demand than a locally produced one. Local firms prefer to 'go abroad' rather than use the indigenous manufacturing industry.

5. Communications and financial institutions were limited to a North/South flow. Until recently, it was impossible to fly between the neighbouring countries of the Gulf; and even now it is difficult to fly between Senegal, Mali, Niger and Upper Volta without going through Paris. Telephone communications were similarly restricted.

6. Since Muslim countries are members of certain currency 'areas' —dollar, sterling, franc—they have to hold all their foreign exchange in one of these currencies. This means that the decline in sterling brings in less 'real' money if those Muslim countries holding it wish to buy goods from the United States or France. Countries in different currency areas find economic union very difficult.

All these barriers have to be removed and trade agreements which give other Muslim countries the 'most favoured nation' treatment must be made.

We feel that the vast sums of money belonging principally to the putative Muslim countries should not be left to incur interest in European banks nor be used to play the money market, but should be used responsibly to promote trade with the Muslim world and finance the development needs of the Muslim countries. Nobody has yet demonstrated satisfactorily why interest should be charged or taken,[10] and as Muslims, being bound to the belief that interest is unequivocally forbidden, we think that any financial aid from these countries should be either:

1. cash grant with no repayment required;
2. interest free loans; or
3. co-operative projects.

Similar aid-types should be used by the principal development bank in the Muslim world, in particular the Arab Development Bank, the Kuwait Development Bank and the recently formed Islamic Development Bank. In fact, the declared aims of the Islamic Development Bank, to provide interest-free loans in accordance with the Islamic *Shariah* to help Muslim countries exploit their natural resources and implement much needed development projects, should also be adopted by other development banks in the Muslim world.

OPEC Aid

The OPEC aid programmes, up to now, have been more than generous. Furthermore, their aid has virtually no strings attached even though they have a strong desire to preserve the now powerful anti-Israel group in the developing countries.

The bulk of OPEC aid has gone to the Muslim countries in general, and the Arab world in particular.[11] Under the pledge made at the Islamic Summit Conference in Rabat in October 1974, Egypt, Syria, Jordan and the Palestine Liberation Organization receive $2,350 million a year. Egypt has also received special loans from Saudi Arabia, Kuwait and Qatar. The aid programmes of Iran have been more liberal and much wider. The Shah's aim is to make his country a leading industrial power, so his aid policies are geared towards the assurance for his country of supplies of raw material, food and markets for its own products. Aid, therefore, has been channelled to Peru and Senegal, Syria, Egypt and Morocco, India and even Britain with this aim. Iran's penetration eastwards will undoubtedly be helped by its stakes in joint shipping companies with India and Pakistan.

Kuwait had a foreign aid programme as far back as 1961. Aid has been channelled through the Kuwait Fund for Arab Economic Development (which has distributed more than £125 million in loans at rates of 3 to 4 per cent interest) and the Arab Fund for Social and Economic Development to which Kuwait contributes around £50 million. Kuwait has also been a major lender to the International Bank for Reconstruction and Development.[12]

It is true that the Arab oil producers' aid is not evenly distributed but concentrated to a high degree in the Arab world. But given the large disparities of wealth within the Arab world and the dictates of Moslem brotherhood and pan-Arab nationalism, it is not surprising that priority should be given to the poorer Arab states. Moreover, in relative terms at least, the oil producing nations are proving to be far more generous donors of aid to the Third World than the industrialised nations of the West. This is the most striking conclusion of a major study of world aid flows for 1974, released by the OECD. Official development assistance from OPEC members came to $2.54 billion in 1974 as against $11.3 billion from the seventeen nations in the OECD's Development Assistance Committee (DAC). However, expressed as a percentage of GNP, the oil producers manage the respectable figure of 1.8 per cent with only 0.33 per cent for the DAC. If all disbursements are taken into account, including portfolio and other investments, the gap is even wider. At $4.75 billion, OPEC's aid represents 3.4 per cent of GNP compared with the 0.77 per cent ($26.4 billion) that the richer nations contribute of their own overall income, well below the theoretical target of 1 per cent. The OECD professed itself 'very impressed' by the speed with which the oil producers have mounted their aid programmes, as well as by the general efficiency with which operations have been carried out. When one recalls that 76 per cent of OPEC's oil revenues go to nine countries with a GNP per head of only $450 a year—less than one-tenth of the $4,600 averaged by OECD members—-OPEC's record is even more impressive. As the *Economist* observed, 'oil exporters are mainly poor but generous'.[13]

By the end of this decade, however, OPEC's concessional aid would not have increased very much. It must decline even if the real income from oil is maintained, which seems unlikely. The priority of OPEC countries is rightly their own development (football teams notwithstanding). Their resources will increasingly be taken by domestic projects. Furthermore, as reserves may dry up in 30 to 50 years there is much interest in profitable investment.

We believe that the 'Arab billions' should be cycled through the developing countries in general, and the Muslim world in particular. On their part, the Muslim countries must be prepared with projects, must negotiate for funds, loans and grants to implement their projects and must provide the understanding that the projects are in fact viable, and where necessary they will be able to repay the loans.

The Muslim world has a rich reservoir of natural resources. Besides 60 per cent of world oil, the Muslim countries produce 70 per cent of the world's natural rubber, 40 per cent of the world's natural jute, 56 per cent of its palm oil, 67 per cent of spices and allied commodities, 30 per cent of black pepper, 80 per cent of Kapok, and 90 per cent of cinchona. There are large deposits of minerals such as iron, copper, tin and bauxite—the last two are in abundant quantities, especially in

Malaysia—manganese, phosphates, chromite, gypsum, limestone, soap-stone, and, in some countries, inexhaustible supplies of natural gas. Even uranium deposits can be found in many Muslim countries in Africa. In livestock and agriculture too the Muslim world stands as an important region of the world.[14] All these reserves await exploitation; and that is where the Arab investments should be. And then there are the rights of other developing countries of the world:

> From the angle of interesting Arab investors in Africa's plentiful raw materials, it is hard to see why African countries should have any difficulties at all. In 1975 Saudi Arabia launched a massive 142,000 million dollar five-year plan which included proposals for setting up entirely new industrial complexes, one at Jabail in the East and another at Yanbu on the Red Sea. The projects listed for these complexes include a steelworks, an iron smelter and a cement plant. Since Africa's reserves of iron ore are said (by the US Treasury) to total twice those of the US and two-thirds those of the USSR, it would seem natural for a country like Saudi Arabia to look across the Red Sea to Africa for a permanent source of supply for that essential raw material. Iron ore is Africa's most widely distributed mineral, and is found in commercial quantities in Algeria, Angola, Egypt, Ghana, Guinea, Ivory Coast, Liberia, Mauritania, Morocco, Namibia, Nigeria, Rhodesia, Sierra Leone, Swaziland and Zambia. But the list of mineral producing countries does not end there. Although up to the present South Africa has seemed to control a wide variety of the markets in African minerals, with its gold, chromite, iron ore, platinum, diamonds, nickel, manganese, asbestos, coal and zinc, it is no longer dominant. Even gold is not a purely South African mineral, being found also in the Congo Republic, Ethiopia, Gabon, Ghana, Kenya, Namibia, Rhodesia, Sudan, Tanzania and Zaire. Diamonds are also widespread through Botswana, the Central African Republic, Ghana, Ivory Coast, Liberia, Namibia, Sierra Leone, Tanzania and Zaire.

Of the minerals that are becoming increasingly vital to industrial output, like bauxite (for aluminium) and copper, Africa has some of the world's most extensive reserves. Copper is currently being mined in Algeria, Angola, Mauritania; Morocco, Uganda and most signifi-cantly in Zaire and Zambia, while bauxite workings are being opened up at a rapid pace in both Ghana and Guinea. Indeed President Sekou Toure claims that Guinea has two-thirds of the world's known bauxite reserves. The foreign companies already exploiting Guinea's bauxite range from American to Swiss, Yugoslav and Russian. In 1974 Guinea hosted the first meeting of the International Bauxite Association, a body of producers modelled on the oil producers' OPEC.[15]

The contribution of the OPEC member countries, however, is not limited to their economic aid programmes, or the investments they have made and could make in the developing countries. The organization has helped the developing countries in many other ways. Most important, it has created a realization of the force that is unity; and now, at least, there is a dialogue on equal terms with the Occident.

Islamic Outlines for Economic Reform

The rises secured by OPEC member countries in oil prices was the first major move for the developing countries and against the present international economic order. OPEC was also the leading force behind the Sixth Special Session of the United Nations General Assembly (first ever to be called by developing countries) which made a unanimous 'Declaration of the Establishment of a New International Economic Order'.

To tip the balance more in the favour of the developing countries, the New Economic Order calls for an internationally agreed link between the prices which the poor world receives for its raw materials and the prices it must pay for the rich world's industrial products.

Secondly, it calls for speedier progress towards 'commodity agreements' to provide developing nations with stable markets at stable prices for stable quantities of raw materials.

Thirdly, it calls on the rich world to restrain its research into synthetic substitutes for the poor world's raw materials — research which is now costing over $1,000 million a year and which has already damaged the market for the economies of developing countries which are dependent on commodities like jute, sisal and rubber.

Last and most important, the New Economic Order urges the developing nations themselves to come together to form producer associations, trade unions of the Third World, to increase their real bargaining power with the rich world and negotiate higher prices for their raw materials.[16]

All in all, these are quite solid suggestions. However, many of the debates regarding the New Economic Order, including those at UNCTAD, have a conspicuous ring of 'international class struggle' about them. As we pointed out earlier, the problems of the Muslim world cannot be solved by substituting one variety of occidentalism for another. Therefore, it is necessary — in fact, vital — that for their New Economic Orders the Muslim societies do not fall prey to any form of occidentalism — capitalism, socialism, or the new emerging structuralism. The task of economic development within the Muslim world is by no means easy. It can only be made worse by treading paths that lead Muslim societies away from Islam and towards occidentalism.

It is true that the task of modern development is not an easy one. There is the enormous backlog of history to be cleared first. There is, in addition, the task of embarking upon a new path, a path still to be conceived and trod on. The World of Islam, in my view, is faced with a much more difficult task than perhaps the other 'worlds'. It carries the stigma of once being a revolutionary, progressive force in the world. Like so many other nations, it just can't play the sedulous ape either to the west or for that matter to the Soviet or the Chinese. If it does, it is afraid of losing its own distinctive stamp and identity among the greatest cultures of mankind. A glorious history, as Islam has, has therefore both pluses and minuses. It must, therefore, struggle to revive its identity and not look for the convenient path of a camp follower. It has to be itself, within itself and of itself. It is a great challenge, nothing short of a 'new birth'.[17]

For the Muslim world the New Economic Order simply means a swing away from the bankruptcy of occidentalism, and towards Islam. As such it would be beneficial to highlight, if only in brief, the economic principles on which the New Economic Order of Muslim societies can be based.

Unlike capitalism and Marxism, the economic principles of Islam do not represent a 'school of thought'.[18] These principles provide a framework somewhat similar to political economy: their function is to discover the laws and analyze the real-life economic situation in the context of a Muslim society with a fully operational Islamic way of life. As such it is difficult to talk about the 'economic system of Islam' outside the context of an operational 'political system' of Islam. The epistemology of Islam, we stated earlier, supports a holistic view of knowledge. One cannot isolate 'systems' from this totality and treat them as if they were independent units, or worse still, 'bits' of knowledge or information. Islam offers a complete world-view: it is not possible to isolate the 'economic segment' of Islam in the hope of obtaining a glimpse of 'economic reality' as presented by Islam.[19]

Neither is it possible to 'Islamize' 'bits of economics' from various occidental economic systems. So what is said about the economic principles of Islam in the preceding pages should be seen in the context of the epistemology of Islam. It is not possible to establish an 'Islamic Economic Order' without establishing the political and social orders of Islam as well.

Within the circumference of Islam the goals of economics are radically different from those of the occidental systems of economics.[20] Here there is no question of agreement either with capitalism or with socialism or even with the now popular Gunnar Myrdal and other structuralists. Islam attempts to solve the economic problem by controlling needs on one hand and distributing resources on the other so that concentration of wealth and power does not take place. If the 'economic system of Islam' concentrates on anything it is the distribution of wealth and reduction of consumption.

The economic principles of Islam emphasize the need to cut down personal consumption as well as to reduce saving and investments, but increase expenditure in the way of Allah. The Qur'an says:

O you who believe! Spend of the good things which you earned and of that which We bring forth from the earth and do not try to spend the inferior part of your wealth in Allah's way, when you would not take it for yourself save with disdain and know that Allah is All-sufficient, All-laudable.[21]

But for those who are niggardly with the bounty Allah has given them, let them not suppose it is better for them, nay it is worse for them, that they are niggardly with what they shall have hung about their necks on the Resurrection Day and to Allah belongs the inheritance of the heavens and earth; and Allah is aware of the things you do.[22]

And again:

Stretch not thine eyes to that. We have given pairs of them to enjoy — the flower of the present life, that We may try them therein, the provision of thy Lord is better and more lasting.[23]

It is obvious, in view of such teachings, that the conventional development policies of economic growth and capital formation are not looked upon by Islam with favour. Islam wants to discourage the growth of consumption, excess profit-motivated production, and excess savings. In other words, Islam solves man's economic problems not by taking him towards materialism but by destroying materialism itself. If you reduce needs, distribute resources equally to all members of society, you strike at the root of poverty and ugly capitalism.

If we think about this a moment we will see that poverty is a function of need. The greater the need of an individual, the greater the amount of material goods and comforts he requires for satisfaction. An income of £2,000 a year in Britain is inadequate for the common man. This is because his needs are many and varied. If these needs were reduced — if the Englishman gave up drinking, gambling, going abroad on holidays, living in semi-detached houses — an annual income of £2,000 may be more than sufficient for him. An increase in economic goods and services has not reduced poverty and deprivation: it has increased it. Those who produce these goods must try to increase the need for these goods, otherwise they will not be able to make a profit. Thus we see that *economic growth does not necessarily reduce deprivation, poverty and injustice.* In most cases in history, *economic growth has increased the level of economic exploitation and injustice.* In the Capitalist world, for example, West Germany and America have a far higher level of economic inequality than Britain but the latter has had a much lower growth rate than the United States and Germany since the Second World War. In the communist world too, Russia's rate of growth has been higher than that of China (since 1965) but the pattern of income distribution in China is more equitable than in the U.S.S.R.[24]

Now, Islam discourages personal consumption beyond a certain limit: beyond the basic needs of the Muslim societies. However, basic needs here include not just food, clothing and shelter — the primary and secondary needs in Maslow's hierarchy of needs — but also general health facilities as well as the medical and domestic care of the sick, the needy and the invalid, general education, and that which may be deemed necessary by the customs and traditions of a particular society. This is the jam the Muslim societies need for their self-development herein; and which will eventually provide them with the jam in the Hereafter.

These basic needs of Muslim societies are met on the basis of two elements of the philosophy of Islam: namely, *tawhid* and *brotherhood*.

Tawhid is the declaration of the unity of God: 'There is no god but Allah.' But it is much more than that. Once you accept the fact that

Allah is the sole Creator of the universe and that you can submit to no one else, you are bound to treat all other men as equal; each man is now a brother to every other man. As far as economics is concerned, this means equality of opportunity and co-operation: '...that natural resources in the universe, such as land, capital, general circumstances such as shortages for reasons of war or disaster as well as laws of nature, all these belong to the whole of society, and all its members have equal shares and right of access to them.'[25] No man has the right to claim a bigger share simply because he cannot or does not create or generate natural power independently.

The principle of brotherhood is really a part of the concept of *tawhid* It is, in fact, *tawhid* in action: equality and co-operation in an operational form. Islam forbids anything that destroys this principle of brotherhood. In particular Islam forbids all forms of interest, gambling, hoarding and speculation. And since *tawhid* manifests itself in economics as equality, man must have free will, to decide what resources to use and develop, and how to plan and direct his material destiny. As such, men's freedom to choose and decide is protected from the tyranny of capitalism as well as socialism. In Islam, therefore, individuals have the right to private ownership.

However, private property in Islam is not an absolute right but a trust from God, and as such one's right to handle one's property is dependent upon the rights and interests of other members of society. For its part, the state is not allowed to encroach upon the liberty of the individual as long as the individual does not violate the rights of the society.

In Islam, ownership of property has quite a different meaning than the accepted conventional, occidental definition. In fact, it amounts to 'stewardship' rather than 'ownership'. An individual is entitled to a share in the total resources of the world as a free gift from nature. The gift is to enable him to utilize and develop his faculties as well as to bring forth goods to his needs as well as those of his neighbours. This is referred to, for the want of better words, as 'private ownership' and it does not imply to all the natural resources indiscriminately. It is limited to those natural resources which can be cultivated by human skills and labour. There are four ramifications of this gift from nature:

1. That ownership signifies only the right to *use* and this ownership can be transferred.

2. That the owner is entitled to 'private ownership' *only* as long as he *uses* it.

3. That the owner who ceases to use his 'wealth' is induced, and in some cases even forced, to part with his idle possessions.

4. That in no case is the owner allowed to charge rent for a free gift of nature from another person who, in fact, has the equal right to its use.[26]

Similarly, work or 'labour' constitutes a different meaning in the framework of Islam. If properly used, labour plays the same part in the development of higher faculties as played by food in that of the physical body. An out-of-work Muslim is in an acute agony: not just because he is in need of an income, but also because he needs nourishment of higher form from disciplined work. See a Moroccan cobbler, observe a traditional Saudi woodworker, watch a Pakistan farmer with his traditional plough—and you will experience an entire cosmos in action. For them, work is not just a form of 'employment' but a creative activity. In their simplicity and non-violence, there is elegance, and there is peace. And this is Islam.

Peace, simplicity and non-violence are also a necessity at the level of the state. In Islam, great stress is laid on ethics and moral values and as such the state must play its part in observing the ethical and moral scene in the country. This includes observance and, if necessary, enforcement of what Muhammed Umar Chapra calls the 'Islamic code of business ethics'. Included in this, he suggests, are five major welfare functions of the 'Islamic State' in the context of economic activity:

(1) to maintain law and order and to safeguard life and property of all individuals;
(2) to enforce the Islamic code of business ethics;
(3) to ensure that the market mechanism works efficiently and to the benefit of all individuals.
(4) to provide physical and social overhead capital: and
(5) to arrange social security.[27]

In addition to these five welfare functions, there are a number of other obligations of society on the state. The most important of these is the collection and administration of *Zakah*[28] —normally translated as the 'poor due'.

Zakah, a sophisticated economic institution, is outlined in the Qur'an, the traditions of Prophet Mohammed and in his actual practice: the *Sunnah*. Muslim jurists and thinkers have produced elaborate treatises of the exact manner of levy and collection of *Zakah*.[29] Our purpose here is simply to emphasize that it must play an important part in the economic policies and planning of Muslim countries. Social security in Muslim countries has usually been instituted with *Zakah*, and it is the responsibility of the state to see that the system of *Zakah* and ancillary measures designed to distribute wealth and resources equally throughout the society function effectively.

The second priority in economic planning concerns the maintenance of equity in the distribution of the resources of nature. It is not enough just to operationalize institutions that distribute resources throughout the society: the state must ensure that this equity is maintained. In other words capital and savings are not allowed to accumulate beyond the desirable, idle possession of resources of nature is condemned, arable land is put to the best possible use, no rent is charged for land, and the Muslim laws of inheritance are operationalized and followed.

This concern for the distribution of the resources of nature and maintenance of equity, will, of course, have a certain impact on economic growth. The connection between distribution and growth is a complete one and depends, amongst other factors, on the nature of society within which income and wealth is being distributed. However, distribution will lead to growth if:

(1)   the initial pattern of wealth and income distribution is so uneven that redistribution increases demand significantly enough to stimulate production;

(2)   there is a rapid increase in population which necessitates the pursuit of economic growth in order to meet the basic economic needs of the poorer sections of the population;

(3)   the state redistributes resources in favour of itself in order to finance investment in an enterprise deemed essential for the propagation of Islam.[30]

The state is also responsible for managing public property —mines, forests, mineral deposits—in the best way that preserves the free character of the endowments of nature and exploit it in a manner that preserves the environmental and ecological balance. The state is also required to manage public utility services which yield profits. Alternatively, these could be managed by a state-approved agency. Also, the state must provide funds for non-profitable public services—roads and bridges, hospitals and clinics, schools and colleges, parks and natural reserves, etc.

In all its planning the state must keep the policy of full employment in focus. This is one of the major characteristics of the 'economic system of Islam'. The primary purpose of this policy is in fact the employment of everyone who needs an income: this would reduce profit and production to second-order determinants and treat people as human beings.

The *Ulema* of Pakistan outlined seven reforms upon which sound economic development could be based. Although directed towards Pakistan, we feel that these reforms are equally applicable in the entire Muslim world as the *minimum* that can be achieved within a short period and without undue effort. We quote in full the summary provided by *The Muslim*:[31]

1. *Interest, Gambling and Speculation.* Interest is the biggest cause of the concentration of wealth in a few hands. Owing to the system of interest, whatever profit is gained from the wealth of hundreds of thousands of people goes into the pockets of a few capitalists who borrow large sums of money to carry on extensive businesses while the people get merely a small sum in the form of interest which too is *Haram*. After obtaining these heavy profits the capitalists start controlling the market. If the Islamic system is implemented this oppressive system would be finished and banking would be carried on, on the basis of sharing profits (*Shirkat* and *Mudharibat*) instead of interest (*riba*).

The second cause of this concentration is gambling. The entire system of insurance is based on it. In addition, the masses also suffer owing to other forms oɪ gambling like puzzles, lotteries and season tickets of various entertainments. In the present system of insurance almost the entire benefit goes to the big capitalists. The poor seldom benefit from it. The Islamic Government should change this system and instead establish associations which would implement insurance according to mutual help techniques. These would not be sullied by interests or gambling and poor masses should benefit from them in an effective way.

Speculation is the third big cause of high prices and concentration of wealth. Owing to speculation, scores of bargains are struck even before the goods come anywhere near the market. Goods worth a rupee sell for three. All this profit goes to the speculators while the masses remain poor. With the prohibition of speculation, prices will definitely go down and the extra profit which now reaches the capitalists will go to the poor masses.

The method of licences and permits in vogue also helps in the establishment of business monopolies. If trade were freed of this oppressive method the prices of many things would go down and the common man would be able to enter the fields of trade and industry with a small sum. Thus the worker of today could own a factory tomorrow.

Lastly, severe penalties should be imposed for hoarding, black-marketing and selling under the counter and hoarders should be forced to bring their stores into the market.

2. *Key Industries under National Control.* The Government should run key industries like steel mills, oil refineries, ship-building, electricity, railways, etc., under its own supervision, and shares in these should only be sold to those whose monthly income is below Rs. 1,000 or whose bank balance is less than Rs. 5,000. The shares of those who have a bigger income or bank balance should be terminated at their appointed time. This method is far more useful than nationalisation of industries because nationalisation does not make industries the property of the poor and instead imposes the sway of the bureaucracy in this sphere.

The present monopolies in industry should also be prohibited and an atmosphere of free competition created so that illegal profiteering may be stopped.

3. *Wage Reform.* The present scale of Government salaries is very unjust and the disparity between different scales is much too great. This disparity should be decreased, the salaries of the high-ups should be lowered and the low salaries be increased.

The wages of workers are very low. In Pakistan the expenditure of a family of five persons is at least Rs. 200 per month but the standard of wages, in comparison, is far less. The wages in various industries range from Rs. 70 to Rs. 125. The labour policy has fixed maximum wages at Rs. 140 but in the present situation this is not satisfactory. A realistic increase is necessary in this. The Islamic Government has the authority to fix such wage rates which may be a suitable return for labour as well as practicable in an industrial society.

4. *Workers and Farmers.* In the worker-entrepreneur relationship the Government could impose a condition that in addition to wages in cash on special performance or after a fixed time or for overtime as a special return instead of paying workers cash bonuses they should be given shares in the ownership of the factory. In this way workers would be able to become share-holders in factories.

For cultivators such rates of share in produce (*Batai*) should be fixed which would be a proper return for their labour as well as reasonable for their essentials of life. As in the case of workers the Islamic Government would have authority to constitute a board on which Government, cultivators and landlords would be equally represented. The exploitation of the cultivators by feudal lords is *haram* and should be forbidden.

In cultivated, unsettled lands the *Sharia* laws of Ihya must be implemented i.e. those cultivators who work the unowned, unsettled lands, themselves, should be given the rights of ownership over them.

5. *Mortgages.* All the present forms of mortgage of lands involving interest should be banned immediately and those lands which are thus illegally mortgaged should be returned and given back to the poor and deserving owners.

A major cause of the accumulation of lands in few hands is that for many years they have not been divided according to the law of inheritance. The Islamic Government should investigate and distribute such lands to their rightful owners. If the Islamic law of inheritance is properly implemented the question of the growth of feudal holdings would not arise.

The methods of transfer of property should be made easier and the free selling and buying of land should be encouraged.

The Government should take measures to provide cultivators with interest-free loans: should give them agricultural implements on easy instalments and arrange for their education and training in agriculture.

The sale of agricultural produce involves many intermediaries and at every stage profit is divided only amongst them, depriving the farmers of their rightful shares. Islam does not like the middlemen system and an Islamic Government should change it.

6. *Nafaqah* and *Zakah.* The Islamic law concerning *Nafaqah* (financial support) should be implemented in its entirety. Besides wife and children, the economic support of certain relatives which Islam has enjoined upon those who have means, should be given a legal shape so as to support orphans, widows, the sick and the incapacitated. If this cannot be done, then the State treasury should provide for them.

A permanent department should be set up to look after *Zakah*. Its working should be as follows:

(a) It should collect *zakah* from those capitalists who have not paid it since the establishment of Pakistan and arrange its distribution among the poor.

(b) Every year it should collect *zakah* on cattle and *Ushr* (one tenth) on lands and distribute it among the poor.

(c) Owners should themselves pay *zakah* on gold and silver and this department should keep a check on rich persons to see whether they have paid *zakah* and *ushr* or not.

7. *Jobs and Housing.* It is the responsibility of the state to provide every person in the country with employment and if in spite of efforts some people remain jobless they should be provided with subsistence allowance till they get jobs.

There are individuals who cannot arrange for their own housing shelter. The Government should set up a permanent housing fund for them and a definite amount should be ear-marked for it in the annual budget.

If the Islamic way of life is established feelings of brotherhood, sacrifice and sympathy will develop and the Muslims of Pakistan will give preference to the pleasure of Allah and Salvation in the hereafter above all material benefits.

This, then, is what can be achieved immediately for the New Economic Order. In all their economic planning the Muslim states should aim at the goals set by *tawhid* and the concept of Brotherhood of Islam. These goals are also the goals of self-reliance and self-sustaining development. Local needs must be met by local resources within the economic framework of Islam. Reliance on others as well as foreign imports to meet the basic needs of a society is an indication of failure and not of success.

NOTES

1. 'The theoretical background is from a Harrod-Domar-Lewis-Rostow doctrine of development in which the main driving force is capital accumulation: and the principal problem is raising the proportion of savings in national income.' Harrod Domar's model assumes 'a cumulative self-sustaining growth' if an economy can show a rate of investment of ten to twelve per cent, with a capital output ratio of three to one and a rate of population growth of two per cent per annum. To set an economy along the path of self-sustaining development, Rostow's 'take off' concept required the fulfilment of following three related conditions:

   '1. A rise in the rate of productive investment from, say 5 per cent of national income.

   2. The development of one or more substantial manufacturing sectors with a high rate of growth.

   3. The existence or quick emergence of a political, social and

institutional framework which exploits expansion in the modern sector and the potential economy effects of the take-off ratio, and thus gives growth. Ziauddin Sardar and Dawud G. Rosser-Owen, 'Science Policy and Developing Countries' in I. Spiegal-Rosing and D. de Solla Price, *Science Technology and Society* (Sage Publication, 1977), ch. 15.

2. See M. J. Fry, *Development Planning in Turkey* (Brill, Leyden, 1971) who applies the conventional model and offers conventional solutions to the development problems of Turkey.

3. Albert Waterston, *Development Planning. Lessons of Experience* (Oxford University Press, 1966), pp. 31-2.

4. See R. Moss, *The Santiago Model*, Conflict Studies 31 and 32 (Institute for the Study of Conflict, London, January 1973), and *Chile's Marxist Experiment* (David and Charles, Newton Abbot, 1974); and Miles Copeland, *The Game of Nations* (Weidenfeld and Nicolson, London, 1969).

5. Tibor Monde, 'From Aid to Re-colonisation', quoted in the *New Internationalist*, 35 (Jan. 1976), p. 4.

6. M. A. Hussain Mullick, 'The Aid Euphemism', *Impact International Fortnightly*, 3 (10), 1 (1973).

7. *New Internationalist*, 35 (January 1976).

8. P.T. Bauer, 'Foreign Aid Forever?', *Encounter* 42 (3), 15-29 (March 1974).

9. Another classic case concerns the part-payment for military equipment by Egypt to the Soviet Union. The payment was made in oranges. The Soviet Union re-sold these oranges for hard foreign currency, thereby gaining financially and strategically. See G. R. Allen and R. G. Smethurst, *Impact of Food Aid on Donor and Other Food Exporting Countries* (FAO, unpub., 1965).

10. For an interesting account of the effect of interest on a society see Anwar I. Qurashi, *Islam and the Theory of Interest* (Ashraf, Lahore, 1967).

11. Three out of thirteen members of OPEC are non-Muslims, namely: Venezuela, Ecuador and Gabon. The Muslim members are: Saudi Arabia, Iran, Iraq, Kuwait, Libya, Qatar, Abu Dhabi, Algeria, Indonesia and Nigeria.

12. All statistics from Zahayr Mikdashi, *The Community of the Oil Exporting Countries. A Study in Governmental Co-operation* (Allen & Unwin, London, 1972).

13. Zahayr Mikdashi and Avi Shlaim, 'OPEC and the Politics of Oil', in Avi Shlaim (ed.), *International Organisations in World Politics Year Book, 1975* (Croom Helm, 1976), p. 161.

14. *Some Economic Resources of the Muslim Countries* (Umma, Karachi, undated).

15. E. C. Chibwe, *Arab Dollars for Africa* (Croom Helm, 1976), pp. 65-7.

16. *New Internationalist*, 32 (Oct. 1975), p. 14.

17. M. A. Hussain Mullick 'Backlog and Development in the Muslim World', *Impact International Fortnightly*, 5 (2), 9.

18. Cf. Khurshid Ahmed, 'Economic Development in an Islamic Framework'. Paper presented at the International Conference on Islamic Economics, Mecca, 5-11 April 1975. See also Sebti Mehdi's analysis of Khurshid Ahmad's paper 'Reformism: A Study in Method, *Suara Al-Islam* 2 (6), 15-30 (June 1976).

19. For a comparison between the economic outlooks of socialism, capitalism and Islam, see K. A. Hakim, *Islam and Communism* (Institute of Islamic Culture, Lahore, 1969); and S. M. Ahmad, *Economics of Islam. A Comparative Study* (Ashraf, Lahore, 1972).

20. For a general introduction to the economics of Islam, see M. A. Mannon, *Islamic Economics Theory and Practice* (Ashraf, Lahore, 1970): M. U.

Chapra, *The Economic System of Islam — a Discussion of its Goal and Nature* (Islamic Culture Centre, London, 1970).

21. The Qur'an 2:267.
22. Ibid., 3:180.
23. Ibid., 20:131.
24. Mehdi, 'Reformism: A Study in Method', p. 15.
25. A. H. A. Abu Sulayman, 'The Theory of Economics in Islam: The Economics of Tawhīd and Brotherhood : Philosophy, Concept and Suggestions for Policies in a Modern Context', in *Contemporary Aspects of Economic and Social Thinking in Islam* (The Muslim Students' Association of US and Canada, Gary, Indiana, 1970), p. 35.
26. S. M. Yusuf, *Economic Justice in Islam* (Ashraf, Lahore, 1971), p. 19.
27. Chapra, *The Economic System of Islam*, p. 40.
28. The exact meaning of the word 'Zakah' is 'growth'. In an alternative sense it connotes 'purification', whence it is used to express a portion of the property bestowed as 'poor-due'. As a form of tax, *Zakah* implies the first meaning of the word, for giving leads to increase of the property in this world and the Hereafter (Qur'an 9:104). Furthermore, the act of giving is also an act of purification. *Zakah* is the duty (*fard*) of every Muslim and as an institution is one of the 'Five Pillars' of Islam. See A. H. Qadri, *Islamic Jurisprudence in the Modern World* (Ashraf, Lahore, 1973), p. 301.
29. See F. G. de Zayos, *The Law and Philosophy of Zakah* (al-Jeddah Press, Damascus, 1960); M. M. Hassein, *Islam and Socialism* (Ashraf, Lahore), pp. 119-77; S. Ahmad (ed.), *Some Socio-Economic Aspects of Zakah* (Pakistan Institute of Arts and Design, Karachi, undated).
30. Mehdi, 'Reformism. A Study in Method', p. 18.
31. Pakistani Ulema Lay down Outlines for Economic Reforms', *The Muslim*, 7 (9), 202-4 (June 1970).

7 A QUESTION OF PRIORITIES: AGRICULTURE OR INDUSTRY?

In all development strategies one is faced with a fundamental question upon which depends the path of self-reliance and self-sustaining growth. It is a question of priority: does agriculture or industry become the cornerstone of development? There is no lack of definite agreement: and the reasons for the choice are more often practical than economic.

Developed countries of the Occident are distinguished by the relatively low proportion of labour in agriculture. This has to be so: only if people are free from the need to produce the basic requirements of life can they work to produce other goods and services. As such, the developed countries have an intense concentration of labour and capital in manufacturing and service industries which is correlated to the standard of living.

This observation is used as the basis for the argument that economic growth of the developing countries can best be expedited by concentrating on the development of industry around urban areas.[1] This has, in fact, been the course pursued by many Muslim countries over the past two decades. Muslim countries like Pakistan, Turkey and Indonesia, eager to gain a certain amount of economic freedom and self-sufficiency, have aimed at rapid industrial growth by investing in heavy capital goods industries. However, the results have not been very encouraging.

Several arguments have been offered in favour of industrialization. Firstly, it is argued that industrial products are not subject to unpredictable and unchangeable factors such as weather. Policies can be outlined easily and plans for the future can be made without any fear. Secondly, industrialization brings beneficial effects on the balance of payments; in the long run it could prove to be an extremely profitable venture. Thirdly, industrialization, it is argued, has a very useful psychological impact on the people of the country. The privilege of working in an 'up-to-date' factory adds drive to the desire for education and technological skills.

Be that as it may, the production problems of industrialization have been immense. The machinery, spare parts, the skilled staff and sometimes even the raw materials have to be imported. If the problems of production have been solved, then a market for the product has to be

105

created—this has often been the decisive factor for the failure of industrialization policies because many developing countries have not been able to sell their manufactures on the world market. On the whole, the road towards industrialization has proved to be an expensive one for many developing countries.

It requires heavy financing and more important, takes out of the economy a great deal of scarce capital in proportion to the value of additional output thus obtained. With only rare exceptions, such as contributions of foreign aid, this places a very heavy burden on the rest of the economy. Restrictions may have to be imposed on the consumption of even essential products, so that these can be exported to pay for the import of equipment. This was done in the Soviet Union when it first started to industrialize. Other sectors of the economy and, in particular, agriculture, may be neglected. In the case of agriculture, it may not only have to forego investment funds for further development but may even be deprived of the minimum resources required to maintain its output. Exclusive concentration on heavy industry at the expense of agriculture tends to lead to a drastic decline in farm production. This happened in the Soviet Union and caused the severe famine of the early 1930's. A similar downgrading of agriculture in the Argentine in the late 1940's and early '50's led to the decline for a long time of the country's main export activities: livestock breeding and grain production. Finally heavy industry, by employing relatively few workers, makes only a minor contribution to solving the paramount problem of unemployment. One must conclude thus that heavy industry offers only limited advantages to the masses in the developing countries on a short-term basis, although the rapid growth of the economy which it fosters will be beneficial to the nation in the long run.[2]

Agriculture, on the other hand, requires little capital, mainly for irrigation, seeds, fertilizers and agricultural machinery. This limited amount of capital properly used could produce handsome economic returns. Unfortunately, agriculture has suffered acutely from neglect, bad planning and too heavy reliance on the Green Revolution.

Emphasis on agricultural production, in contrast to industrialization, augments the incomes of the rural populace. There is, however, a problem in mobilizing this additional income for investments. Here only carefully designed taxes can provide incentive for increased output as well as private savings and additional income needed for investments.

It is not easy to say that emphasis on agriculture will lead to increase in employment. This depends on the extent of modernization as well as the system of land tenure of a country. On modernized farms where land is concentrated in large holdings, mechanization increases movements from rural areas to urban slums.

We would argue in favour of a balanced approach: one that re-elevates agriculture to its proper place and concentrates on light rather than heavy industry. The accent must be on increasing employment and eliminating poverty. Emphasis on one to the detriment of the other produces the kind of lopsided economies and societies that are all too obvious in the Muslim countries. Complete reliance on heavy industry produces concentration of 'modernity' in a few cities while the rural areas remain poverty-stricken, and unemployment continues to mount. Too much emphasis on agriculture deprives a country of much-needed local industry.

In our opinion, there is a need for balance: a need to escape from fashions and fads which may be appropriate for other societies, a need to choose those fields of agriculture and industry which can be developed most advantageously, a need for more employment and less mechanization. A balanced approach would be an Islamic approach.

The undue emphasis on heavy industry in the Muslim countries over the last two decades has led to neglect, and as a consequence, to poor performance of agriculture. In some countries agriculture has been allowed to lapse into such a state of stagnation that only drastic measures would re-elevate it to its due status in the countries' economy.[3] In such areas agriculture must be given top priority until the return to normalcy. There are at least five reasons for this:

1. Since the Muslim countries are, by and large, agriculture or mineral producing, an increase in the prosperity of the rural population will increase both savings and demand, thereby stimulating growth.

2. The raw goods that agriculture provides feed industry and the export-orientated sectors. So an increase in agriculture would result in an increase in foreign exchange and employment. Furthermore, the prices of raw materials are increasing in the world market and will increase even more as developing countries form themselves into organizations of raw material exporting countries. As agriculture, when properly planned and managed can create employment, so it can trigger off a migration in reverse from cities to the rural areas.

3. Improvements in agricultural productivity make widespread economic growth easier and sustainable by transforming a subsistence economy into a market economy.

4. 'A rapid growth of agricultural productivity is important, as it enables food supplies to be available at relatively lower prices. The non-agricultural sector then requires less of its income to purchase food, so increasing the effective demand for output of the non-agricultural sector. This in turn increases the profitability of an expanded output in the non-agricultural sector and encourages entrepreneurs to invest there. Concurrently expansion of the non-agricultural sector will increase the availability of job opportunities in that sector, both for the urban population and the labour released from the rural areas. Furthermore, relatively declining food costs imply higher real incomes thus reducing the pressure to increase wage earnings. This maintains or increases the profitability of investments in the non-agricultural sector. Lower food prices also reduce political discontent.'[4]

5. Agriculture does not compete with industry — the two sectors are complementary. In Muslim countries it is futile to rely on any kind of industry without strengthening the agricultural base. The industrial revolution of Britain was based on an even more successful agricultural revolution.[5] A strong agricultural sector provides a strong base for industrialization.

It has been convenient to conceal the mistake of overlooking the importance of agriculture in development, and blame the poor performance of agriculture on the adversity of nature, on the perversity of farmers, and on the fecundity of man.

A spell of bad weather, a season of drought, a sequence of heavy rainfall, and nature comes under attack. But, 'accuse not nature', as John Milton has said, 'she hath done her part; do thou but thine'. Nature did her part in Bangladesh, in Pakistan, in the Sahel. Men, on the other hand, still plan to act. The variability of weather should, of course, be an integral part of any normal expectations with regards to agricultural production. The droughts will not stay for ever, neither will floods come every year, nor will it rain forever. Calamities of nature will come and pass; there will be good years when the output will be more, much more than enough. The adversities of nature have always been there, and always will be there. They do not account for the poor performance of agriculture.

In the minds of many who shape economic policy, farmers are ever so perverse. When a national economic plan calls for more agricultural production, farmers fail to respond; when instructions are issued to shift from wheat to corn, they fail to produce enough of either crop; when given the command to make a big leap forward, they step backward; and when they are heavily subsidized to reduce the acreage of particular crops, they proceed to increase the yield to more than offset the reduction in acreage. It has been convenient to believe that farmers, especially in poor countries, are loafers who prefer leisure to doing the extra work that will increase production; are squanderers who will not save for investment with which to increase agricultural production, and are inefficient in using the resources at their disposal. Thus these poor, lowly farmers are to blame. But farmers are not perverse in their economic behaviour. If there is perversity, it will be found in the minds of those already mentioned, in what they behold in agriculture, and in national economic plans that fail to provide economic incentives for farmers.

It is now fashionable to attack the fecundity of man as the culprit, as if it were to blame for the poor performance of agriculture. The excessive growth in population is indeed a serious matter, for surely it has major adverse social and welfare effects in what can be done to improve health facilities, to enlarge cultural opportunities and to provide schooling; and it can be a heavy drag on economic development. This excessive growth in population, of course, also increases the demand for food; nevertheless, the rapid growth of the population is not responsible *per se* for the poor performance of agriculture.[6]

The real culprit is neglect. Agriculture has performed badly in the Muslim countries because it was deprived of vital economic opportunities which were diverted towards heavy industry. When attention was given to agriculture it was usually in the form of blind rural reforms and introduction of imported, out-of-place mechanization.

The neglect of the agricultural sector was nowhere more apparent than in underpricing of farm goods. Why Muslim countries, and indeed many developing countries, engaged *en masse* in this absurdity is not quite clear. But what is clear is that low prices for agricultural goods has meant a low incentive to increase production. Capital can safely be invested elsewhere for quicker, greater returns.

Import substitution, while theoretically saving foreign exchange, made goods more expensive. When the use of farm machinery became necessary for increased production it was out of reach of the pocket of the average farmer. So only the rich could benefit.

In general farm machinery is rarely available, and when it is, repair facilities and the supply of spare parts is notoriously inefficient. In some countries, production of these goods is restricted to the public sector; and where it is not, competition is not as sharp as it should be and areas of monopoly exist. Fertilizer too has remained out of reach of the average farmers. For these and other reasons, while prices of consumer goods have been rising up to now, relative to the prices of farm products, there has been a continuous decline in the living standards of the farmer relative to his city-dwelling cousins.

At the same time, it has become harder and harder to work the land. Water shortages have affected irrigation. In dry lands of the Muslim world with low rainfall and an acute lack of secondary irrigation, farming is a struggle all the way. While many governments recognize the problem, no help has been rendered to the farmer. Not much research has been done on his specific problems.[7]

Farming in the Muslim world, as has often been said, is not just an economic activity: it is an active way of life. In this life-style everything fits in its place and flows gracefully. Introduce an alien idea or piece of machinery which does not suit this life-style and you will upset the entire system. It is possible to put constraints on this system to increase 'productivity' up to a point: beyond this threshold the system falls to pieces.[8]

The farmer is considered simply as a producer who must cut his costs and raise his efficiency by every possible device, even if he thereby destroys—for man-as-consumer—the health of the soil and beauty of the landscape, and even if the end effect is the depopulation of the land and the overcrowding of cities. There are large-scale farmers, horticulturists, food manufacturers and fruit growers today who would never think of consuming any of their own products. 'Luckily,' they say, 'we have enough money to be able to afford to buy products which have been organically grown, without the use of poisons.' When they are asked why they themselves do not adhere to organic methods and avoid the use of poisonous substances, they reply that they could not afford to do so. What man-as-producer can afford is one thing; what man-as-consumer can afford is quite another thing. But since the two are the same man, the question of what man—or society—can really afford gives rise to endless confusion.

There is no escape from this confusion as long as the land and the creatures upon it are looked upon as nothing but 'factors of production'. They are, of course, factors of production, that is to say, means-to-ends, but this is their secondary, not their primary, nature. Before everything else, they are ends-in-themselves; they are meta-economic, and it is therefore rationally justifiable to say, as a statement of fact, that they are in a certain sense sacred. Man has not made them, and it is irrational for him to treat things that he has not made and cannot make and cannot recreate once he has spoilt them, in the same manner and spirit as he is entitled to treat things of his own making.[9]

This confusion is largely responsible for the present state of agriculture in the Muslim world. Once you have played havoc with their life-style, the farmers can do little but migrate to cities for employment. They must eat. And the cities carry the smell of food.

The current distribution of labour force in the Muslim world we have estimated to be: agriculture, 45-60 per cent; industry, 12-19 per cent; tertiary, 21-32 per cent. In contrast, the figures for the Occident are: agriculture, 3-15 per cent; industry, 33-46 per cent; tertiary, 48-61 per cent.[10]

If mechanization continues in the agricultural sectors of the Muslim countries, we believe that unemployment will continue also. The unemployment situation will be further augmented by the commitment of many Muslim countries to heavy industry. In Algeria, where the policy of 'walking on two legs' is pursued, traditional Arab agriculture covers about 75 per cent of all arable land and employs nearly 80 per cent of the rural labour force. In contrast, socialist agriculture established in large mechanized farms, covering 25 per cent of arable land employs only 1 per cent of the labour force. As more and more of traditional Arab agriculture is taken over by 'modern' agricultural techniques, unemployment rises functionally.

The Green Revolution

Most of the modern techniques of agriculture were introduced in the Muslim countries in the form of the Green Revolution. Designed in the Occident, the Green Revolution was introduced as a package solution for the countries of 'one-crop' economies.[11] Considered a major contribution to the battle for expanding grain production in the food deficient countries, the Green Revolution focused on the development and international dissemination of high yield variety seeds of wheats and rices. Even up to the beginning of this decade these seeds were talked about in breathless wonder. Thus Lester Brown:

> The impact of the new seeds on production levels is a result of their prairie-fire-like spread wherever conditions are suited to their use. Their widespread adoption in turn relates to their remarkable capacity for doubling yields of the varieties they replace. Both the wheats, developed in Mexico by the Rockefeller Foundation, and the rices, developed by Ford and Rockefeller jointly in the Philippines, are of short stature. It is their short, stiff straw, enabling them to respond to up to 120 pounds of nitrogen per acre, which distinguishes them from the traditional thin-strawed varieties which begin lodging (falling down) at about 40 pounds of nitrogen.
>
> In addition to their exceptional yield capacity, the new seeds mature earlier and adapt to a much wider range of latitudinal and seasonal variations than do indigenous varieties. By maturing in 120 days instead of the 150 to 180 days required for the varieties they replace, the new rices open up new opportunities for multiple cropping.
>
> Countries traditionally in food deficit are now using the new seeds and becoming self-sufficient, some actually generating exportable surpluses. The Philippines, the first country to use the new rices on a commercial basis, has ended half a century of dependence on rice imports, becoming a net exporter. Pakistan, as recently as 1968 the second-ranking recipient of United States food aid, has sharply reduced its dependence on food imports and is expected to be self-sufficient in both wheat and rice in 1970. Food imports into India are now less than one-half those of the food crisis years of 1966 and 1967.[12]

During the mid-1960s and early 1970s some success was accomplished by the Green Revolution. Pakistan,[13] Indonesia, Morocco, Egypt, Iran — all nearly doubled their yield. The momentum, however, has now been lost. Tables 2 and 3 give some details of the use of high-yield varieties of wheat and rice by some Muslim countries.

In addition to high-yield seeds, the Green Revolution involves a sophisticated form of agriculture which needs large amounts of water, fertilizer and pesticides. As an example it requires four to seven times more water per acre to achieve the crop yields that characterize the Green Revolution compared to traditional practices which use the old variety of seeds. The acute scarcity of water in many Muslim countries

Table 2: Use of high-yield varieties of wheat by some Muslim countries

Country	Proportion of planted area with high-yield varieties (percentages)				Area planted to high-yield varieties, 1972/73 (thousands of hectares)	Yield (quintals per hectare)					
	1969 /70	1970 /71	1971 /72	1972 /73		Average 1960/61 1963/64	1970 /71	1971 /72	1972 /73	1973 /74	Average 1970/71- 1973/74
Afghanistan	6	11	13	18	450	9.6	8.2	8.9	11.7	12.3	10.3
Algeria	2	7	15	28	600	7.1	6.1	5.9	6.1	5.1	5.8
Bangladesh	8	11	12	18	21	6.0	8.8	8.9	7.9	8.3	8.5
Egypt	–	–	–	90	–	25.4	29.5	30.5	31.0	32.2	30.8
Iran	2	6	6	7	298	8.0	9.0	8.3	9.4	9.3	9.0
Iraq	10	9	45	23	457	5.2	5.2	5.9	7.6	4.0	5.7
Jordan	4	4	5	10	–	3.6	3.1	6.7	9.6	4.0	5.9
Lebanon	4	12	19	31	20	6.5	8.2	7.5	10.2	4.7	7.7
Morocco	3	5	10	13	294	6.3	10.0	11.7	11.7	8.7	10.5
Pakistan	43	52	57	56	3,339	8.2	11.7	10.8	11.9	13.1	11.9
Syria	–	4	7	21	180	6.1	7.1	6.9	10.0	6.5	7.6
Tunisia	7	11	6	10	99	4.1	6.0	6.3	8.0	7.9	7.1

Source: D. C. Dalrymple, *Development and Spread of High-yielding Varieties of Wheat and Rice in the Less Developed Nations*, United States of America, Department of Agriculture Economic Report No. 95 (Washington, D.C., July 1974).

Table 3: Developing countries of South and East Asia: rice area and yields of some Muslim countries, 1972-1974

Country	Area (millions of hectares)				Percentage planted with high-yield varieties				Yields (quintels per hectares)					
	1971	1972	1973	1974	1969	1970	1971	1972	Average 1966-70	1971	1972	1973	1974	Average 1971-74
Bangladesh	9.5	9.5	9.9	10.1	3	5	7	11	16.9	16.5	16.3	18.6	18.3	17.4
Indonesia	8.5	8.6	8.5	8.5	10	11	16	18	20.4	23.2	21.9	26.7	28.7	24.9
Malaysia (West)	0.6	0.6	0.6	0.6	26	31	36	38	26.9	29.9	29.5	28.8	28.9	29.3
Pakistan	1.5	1.6	1.5	1.6	30	36	50	43	18.9	14.7	14.2	24.3	23.3	19.1
Philippines [1]	3.2	3.1	3.4	3.5	44	50	56	56	14.9	15.7	16.3	16.3	16.9	16.3

1. For comparison.

Sources: D. G. Dalrymple, *Development and Spread of High-yielding Varieties of Wheat and Rice in the Less Developed Nations*, United States of America, Department of Agriculture, Foreign Agricultural Circulars, *Rice*, Nos. FR 1-73 and 2-74 (Washington, D.C.).

produced its own problems.[14] Furthermore, the fertilizer which in fact is the basis of the Green Revolution has been in tight supply for some time. And what is on the market is out of the economic reach of many Muslim countries.

Although the output, up to a few years ago, was high in quantity, it has always been low in quality. The fat and carbohydrate content of the new grain is high while the protein content is very low. The hybrid grains produced by the Green Revolution are highly susceptible to insect pests and plant diseases. This has forced an increase in the use of chemical pesticides. Also a dangerous trend towards crop uniformity has developed in many Muslim countries. Acres of land formerly used for nutritious vegetables, fruits and beans have been used in the rush to produce higher grain yields.

The side effects of the Green Revolution have often been disastrous.[15] Millions of acres in Pakistan, Egypt and Turkey have been transformed into veritable salt deserts because of the excessive use of irrigation techniques. Irrigated acres have also become fertile ground for parasitic diseases such as malaria and schistosomiasis. The irrigation network in the Aswan Dam, for example, has completely destroyed the local sardine industry, reduced the productivity of the Nile in general and caused schistosomiasis on a wide scale.[16] In addition the Green Revolution has done serious damage to the biosphere: pesticides such as DDT with high toxicity enter and accumulate in food chains, killing birds and fish by affecting their reproductive processes.

The social effects of the Green Revolution have been just as serious. Its monocultural approach has caused dislocation and dispossession of the rural people. Its extreme capitalist orientation encourages the emergence of farmers who produce individually and enter into a monetized mode of production. The often uneven distribution of economic benefits widens the gap between those people with enough resources to capitalize on the new variations and those who are poor. With its heavy reliance on fertilizers, the Green Revolution has done more for the profits of the agricultural business complexes (chemicals, machinery, etc.) than for the long-term interests of the rural population in the Muslim world.

Any agricultural strategy based on the Green Revolution suffers from three major problems:

1. Prices of primary goods have been decreasing, with few notable exceptions, whereas those of manufactured goods have been rising. More, therefore, has to be exported to pay for the imported

machinery needed for the Green Revolution.

2. Increasing agricultural productivity has meant increasing output per man rather than output per acre. This naturally leads to unemployment and forced exodus of rural workers to urban centres. However, to operate fully and efficiently, smallholdings have to be enlarged for the use of machinery. This has often led to land reform policies which have been detrimental to poor farmers. Even worse results are obtained from policies of enforced collectivization such as those followed by Lenin and in the Red River Delta in North Vietnam.

3. Agriculture is much more than an 'economic activity'. For some it is a 'way of social existence'. To persuade villagers to accept certain new techniques virtually amounts to asking them to change their life-style. This may be easy to ask, but it is quite difficult to achieve.

In view of the overall failure of the Green Revolution, and the often disastrous side effects of 'modernizing' agriculture we would suggest a more balanced approach to agriculture which uses the best of traditional techniques coupled with the new techniques of intermediate, indigenous technology. We will have more to say about indigenous technologies in Chapter 8.

We believe that revitalization of agriculture in Muslim countries goes hand in hand with policies for rural development.

We believe that success in agricultural growth can only be achieved by ecologically adapted and economically viable agricultural technology which involves a continuous adaptation to locally available resources as well as positive response by cultural, economic and political forces. The process of agricultural development must be induced by the farmers themselves in conjunction with agribusiness and administrators. Needless to say, the non-agricultural sector plays an important role in this process, as the suppliers of technical input and providers of technical assistance as well as marketing facilities. As such, local industry must be able to feed the agricultural sector with cheaper services, as well as having the capacity to generate a continuous sequence of technical innovation in farming techniques which should not just develop the agricultural sector but also increase the demand for inputs supplied by the local industry.

Growth of agriculture on these lines would mean that development is achieved by relying solely on local resources. This type of self-reliance requires a co-ordinated approach as well as development of

(1) an integrated food and agriculture policy; (2) planning for an integrated development of a national food industry; (3) adequate government stockpiling, particularly emergency reserves of perishable foodstuffs; (4) adequate education and training facilities of food and agricultural technicians at all levels; and (5) development of small-scale rural industries. The last pre-requisite is particularly important: without an adequate backup industry the agricultural sector would not be able to receive the push it so acutely needs.

The Road to Industrialization

The two types of industrialization programmes currently being implemented in the Muslim countries include (1) those designed to substitute domestically produced commodities for imported goods; and (2) those industries designed to increase exports and thereby increase foreign exchange earnings. Both rely on heavy industry and have several shortcomings.

The industries designed to substitute imported goods have to be protected from longer established, more efficient foreign competitors. This is normally achieved by introducing import tariffs. However, there is a real danger of retaliation from the occidental countries: a large segment of the Occident does not take kindly to the developing world protecting its nascent industries. Furthermore, the goods produced are often those in which the country does not have a comparative advantage. Many products are designed to suit occidental tastes and pockets. The modes of production are distinctively occidental and do not take into account indigenous raw materials and skills. The end result is often that of a liability rather than an asset. Even in the long run, there are little or no savings in foreign exchange.

Consider the case of the Algerian steel industry as an example. The Algerian Government has decided to give absolute priority to industrialization. The policy was based on Soviet development patterns which were designed to be imitated. In many ways Algeria was fortunate to have abundant deposits of iron ore and it was thought that production of iron and steel made good economic sense. In its initial stages the costs of constructing steel mills were particularly high and the capacity of the new plants was substantially in excess of the domestic demand. A substantial part of the output had to be exported to the highly competitive world market, which not only reduced but quite offset Algeria's profitability. In the end, the indirect costs in foreign exchange of producing one ton of steel became greater than the savings in foreign exchange, and consequently indigenous steel is more expensive than imported steel.[17]

Sometimes strategic interests may override economic ones and it may become necessary to establish heavy industry for domestic as well as export purposes. But where even raw materials have to be imported, the strategic importance of the industry is devalued.

The policy of concentrating on the export sector may prove beneficial for development in the long run, but the short-run effects almost always range from bad to disastrous. Heavy industry, as we know, is capital-intensive: establishing heavy industry by transfer of technology often results in the country being mortgaged to the last hut. Local management often fails to operate at the level of sophistication that is necessary,[18] and the ever-present problem of unemployment becomes more and more acute. As back-up facilities of supporting this high technology industry are non-existent, the plants waste valuable production time for want of spare parts which have either been delayed or for which the foreign exchange is not available. Bureaucratic inefficiency further cripples any hope for full production.

Of course, it is not enough just to produce: one must also sell one's products. Successful exporting depends on both quality control and marketing, and at both levels the Muslim countries are inept. Marketing in particular suffers from the fact that larger exporting industries are usually government owned, and are subject to the lethargy most nationalized industries are renowned for.

The most often used way of solving the marketing problem is to arrange for some competent marketing firms to take care of the process and to have either the right trained personnel in quality control or expatriates until native technicians can take over. This, however, is not a road which leads to self-reliance.

At the base of policies for heavy industrialization is national pride. Establishment of heavy industry in many Muslim countries is somewhat akin to the American national goal of landing a man on the moon. Both policies have little to do with the social needs of the populace. And what could be more beneficial for national pride than conspicuous technology?

Light industry, in contrast, is much less conspicuous, less spectacular, less prestigious, and, as such, is not considered to be an indicator for national progress. Nevertheless, we feel that concentration on light industry is a much saner way towards the goals of self-reliance and self-perpetuating development. In particular, light industry needs less capital, more labour and provides reasonably quick returns. These characteristics make light industry less of a burden on other sectors, and often a country can pursue healthy agricultural development patterns at the same time as promoting light industries. More than that, light industries provide the back-up so essential for the development of the agriculture sector.[19] As these industries operate at a low level of technology, they are less dependent on foreign spare parts and easier to manage for the local managers. As such they provide an ideal base for the development and use of local resources, manpower and skills.

Moreover, light consumer industries can be more easily orientated towards local needs and requirements, producing what is important and using what raw material is available locally.[20] Although these industries cannot produce the large amount of capital promised by heavy industries in the long run, they can, nevertheless, accumulate and facilitate the accumulation of capital which provides a slow but steady rate of growth. This leads to a more equitable distribution of wealth. Because of the low level of technology employed in many light industries they are less dependent on large towns — thus, industry can be taken to rural areas.

It is necessary, in our opinion, that Muslim countries pursue policies which reduce imports and encourage the development of local light industry. This is to say that emphasis should be placed on reducing imports, but at the same time certain export-orientated policies can also be followed. In particular, those policies which direct exports to other Muslim countries should be encouraged. Whereas export-orientated industries are heavily influenced by factors outside the country, industries aimed at decreasing imports are less so. Governments can do little about foreign competition, but it is not difficult for them to enforce policies that ensure that import-orientated industries neither become stagnant nor acquire monopolistic tendencies. Furthermore, import-orientated industries are at least assured of a market where they can operate at full capacity. Over a long period, import-orientated industries can provide both the capital and the expertise needed for an efficient export sector. Until that time, the necessary foreign exchange can be accumulated through the export of raw materials and development of associated light industries. Some Muslim countries, after all, need a strong export sector, if only to pay off their large debts.

For the achievement of the goals of self-reliance and self-development, we must stress that it is necessary — in fact, vital — to concentrate on the development of local technological skills and light industry. Not just a decentralization of industry is needed, but also a positive swing away from the reliance on foreign technology and 'imported know-how'. This requires us to be more critical of conventional technology and approach it from a new, fresher angle.

NOTES

1. See, for example, Albert Hirschman, *The Strategy of Economic Development* (Yale University Press, New Haven, 1958), pp. 98-110, where he argues that heavy industries, notably iron and steel, are the most powerful stimulant for economic growth.

2. Paul Alpert, *Partnership or Confrontation: Poor Lands and Rich* (Free Press, New York, 1973), pp. 48-9.

3. See, for example, E. A. Bawany's description of the state of Pakistan's agriculture, *Revolutionary Strategy for National Development* (Muslim News International, Karachi, 1970), ch. 4.

4. David Metcalf, *The Economics of Agriculture* (Penguin, London, 1969), p. 75.

5. See E. Mingay, *The Agricultural Revolution* (Batsford, London, 1967).

6. T. W. Schultz, 'What Ails World Agriculture?', *Bulletin of the Atomic Scientists*, January 1968, p.29.

7. For the little research that is being done on agriculture, see B. F. Johnston's survey of largely theoretical research 'Agriculture and Structural Transformation in Developing Countries: A Survey of Research', *J. Econ. Lit.* 8 (2), 369-404 (1970).

8. 'Agriculture is a systems problem, it will perform effectively only if a whole range of interacting conditions are satisfied.' M. P. Milliken and D. Hosgood, *No Easy Harvest* (Little Brown, Boston, 1967), p. vii.

9. E. F. Schumacher, *Small is Beautiful. A Study of Economics as if People Mattered* (Blond and Briggs, London, 1974), pp. 87-8.

10. Based on M. Longelle, 'Labour Productivity in Agriculture and the Balance Between the Three Main Sectors of the Economy in Developing Countries', *Revue Europeenne des Sciences Sociales* 10 (26), 223-7.

11. For a general introduction to the Green Revolution, see S. Johnston, *The Green Revolution* (Hamish Hamilton, London, 1972).

12. Lester Brown, 'The Green Revolution', in Barbara Ward *et al. The Widening Gap* (Columbia University Press, 1971), p. 128.

13. See Leslie Nutty, *The Green Revolution in West Pakistan. Implications of Technical Change* (Praeger, New York, 1972).

14. The UN Food and Agriculture Organization has estimated that some nations will probably experience acute water shortages by the end of the 1970s without any further increases in the use of irrigation.

15. J. Milton gives some examples in *Careless Technology; Ecology and International Development* (Tom Stacey, London, 1972).

16. The controversy over the Aswan Dam has been going on for years now, see J. Rzoska (ed.), *The Nile. Biology of an Ancient River* (W. Junk, The Hague, 1976), who presents both sides of the argument. A summary article appears in *Nature* 261, 444-5 (1975) from where the self-explanatory table below is taken:

Nile Waters during an average year
with and without the Aswan High Dam (in million m^3/year)

	With	Without
Water amount available in Sudan	85,000	85,000
Allocation to Sudan	− 18,500	− 13,000
Evaporation losses in new lake	− 11,000	—
Water passing Aswan	55,500	72,000
Irrigation of Upper Egypt	− 24,300	− 18,800
Recirculation gains	+ 5,300	+ 8,000
Water passing Cairo	36,500	61,200
Irrigation of Delta	− 36,500	− 28,200
Discharged into Mediterranean	—	− 33,000
	0	0

17. Alpert, *Partnership or Confrontation?*, pp. 161-2.
18. K. J. Shane points out that four types of 'process aid' from the Occident can improve the management technique in developing countries
    (1) to overcome obstacles to efficient operation of management development;
    (2) to make these processes shorter and easier;
    (3) to help out particularly difficult stages of the process; and
    (4) to improve the efficiency of the use of existing resources.
    See *Quest:* Journal of the City University, 29 (Spring 1975).
19. M. Bandini describes some small and medium size industries which can thrive in rural conditions. See his 'There is space between Farm and the City', *Cares* 3 (5), 51-4.
20. E. Kung makes much the same point and argues that if developing countries produce these goods for which they are particularly suited, regardless of whether these export goods are industrial raw materials or farm produce, the whole controversy between industry and agriculture will lose much of its intensity. See his 'Agriculture vs Industrial Development in LDCs', *Inter-economics*, 6, 177-80 (1972).

8 IMPORTED KNOW-HOW OR TECHNOLOGICAL SELF-RELIANCE?

There seem to be as many definitions of technology as there are technologists. We find definitions of technology ranging from the general use of knowledge to 'the science of industrial arts',[1] 'tools, including machine, but also including such intellectual tools as computer languages and contemporary analytic and mathematical techniques',[2] 'the systematic applications of scientific or other organised knowledge to practical tasks'[3] and even 'the grammar of the future'.[4]

To us 'technology' is a social impression rather than a definition. The closer you get to the impression the more difficult it becomes to define it. After Dickson, we have used the word in a 'broad social sense in a way that suggests how the relationship between technology, machine and technique might be seen or roughly equivalent to that between language, words and speech'.[5]

One of the more public effects of having an occidentalized elite taking the political decisions involved in science policy, and indeed in having occidentalized scientists and technologists who are themselves willing to participate in such schemes even if they are not actually rooting for their adoption, is Conspicuous Technology.[6] This is a sort of international 'keeping up with the Joneses' or 'gamesmanship'. Like the ears of Midas, Conspicuous Technology is difficult to hide. It clearly illustrates the problem of who is taking what decisions about science policy, and whether they are really appropriate for the domestic needs of the country. A clear example, in our opinion, of such profligacy is the Indian Aerospace Programme or the Atomic Energy schemes of many countries. The Indian Atomic Energy scheme also has the element of Conspicuous Technology taken to an extreme with its development of an atomic bomb: this move can only be seen by its neighbours as aggressive and expansionist, and can lead to the destabilization of the area of South Asia and add a further element to the already destabilized South-East Asian and Middle Eastern spheres, with the consequent arms race that this produces. And yet few of these countries have sufficient domestic resources to keep up their current occidentalized development programmes without competitive, escalating arms programmes being added to the bill. Parallels to this can be found all over the globe; it is a major problem of developing countries that such prestige projects are embarked upon to the neglect of more necessary but less prestigious ones. As William Shakespeare has said, 'The fault lies not in our stars but in ourselves that we are underlings'. Here the real problem would appear to be less inappropriate development programmes and rather more inappropriate decision-takers.

The Transfer of Technology

Conspicuous Technology is largely the result of technology transfer — a typical form of relationship between the rich and the poor. Technology transfer occurs when a firm in an occidental country claims to have what the Muslim country *thinks* it wants or needs. Often it is the owner of the technology who persuades the Muslim country that it really needs or wants that piece of technology. As there is a euphoric feeling in the Muslim world about the wonders and benefits of modern technology this does not prove difficult for the occidental salesman: demand for technology, natural or artificial, is always there. The negotiators of the Muslim world even lack the expertise to ascertain a proper price for technology purchased off the shelf, let alone the knowledge to break a package into parts, some of which, as it turns out, could have been handled domestically. There is no question of their insisting on the design of low cost systems tailored to their needs. As a result, along with the package come the managerial and technical skills. A system of patents and licences neatly ties the Muslim countries to a multinational corporation. From now on all the spare parts, and ancillary material must be purchased from the licensee — never mind the fact that most of it costs up to 80 per cent more than what is available on the international market. This is why American earnings on foreign licensing rose from $650 million in 1960 to $1,858 million in 1969.[7] The poor country, on the other hand, becomes even poorer.

Patents and licences involve not only material objects and processes but also 'know-how'. This latter notion is almost impossible to define, and, as tends to happen when commercial interests conflict over meanings and definitions of key terms, the legal profession develops a speciality to meet the difficulty and to reap the benefits of the confusion sown in the first place mainly by colleagues originally responsible for framing the laws. Indeed, patent lawyers are a growing body in most countries — they have work to do in poor as well as rich societies — and even the most cursory glance at *The Encyclopedia of Patent Practice and Invention Management* suffices to confirm one in the view that the whole matter is fraught with difficulty and ambiguity.[8]

But even the 'know-how' imported is of a very poor quality. Over the past two decades, the multi-nationals have consistently withheld their sophisticated technology from the host countries, and the technology they have transmitted has often turned out to be obsolete, overpriced and inappropriate. And as the parent company is responsible for research there is no attempt on their part to adjust the imported package to the local needs.

The sold package, therefore, almost always tends to support the interest of capital and technology suppliers.[9] The overall contribution to development of the importing country is usually feeble and sometimes negative.[10]

By and large conventional occidental technology has evolved along lines appropriate to the conditions and circumstances of the Occident. It has inherited the occidental values and norms, it is an expression of occidental culture.[11] It is capital intensive, labour saving, and production orientated. Such a technology is quite inappropriate for Muslim countries where (1) labour is abundant or superabundant; (2) capital is extremely rare; and (3) there is often an acute shortage of skilled labour and management. (The first two of these conditions do not apply to the OPEC nations where now capital is plentiful, but labour is acutely scarce). An example of inappropriateness of conventional technology is provided by Boon.[12] He compared the total cost of making wooden window frames in the developed countries and in the developing countries. He found in the labour expensive country two special-purpose machines, one performing four-sided planing and moulding, and the other a double-sided tenoner, with the cheapest calculations for output capacities in excess of 50,000 units per annum. In the developing country with cheap labour and dearer capital, however, the whole process was uneconomical unless the capacity exceeded 450,000 units per annum. Up to an output of 64,000 which encompasses the great majority of carpentry workshops, the lowest unit cost would be achieved by using single ended tenoners, with considerably lower investment per worker.

Production-orientated technologies are designed mainly for use in large-scale plants. Thus, certain basic requirements must be fulfilled before they can be used with economic benefit in the Muslim countries. For example, production-orientated technologies require well-developed large marketing facilities and distribution channels, continuous (24 hours) operation, sophisticated management and well-disciplined workers. However, it could prove disastrously uneconomical if markets are small, scattered, seasonal or fragmented; if distribution channels are not organized; if management lacks skill; or if workers are not disciplined or used to night shifts. In such circumstances, imported high technologies cannot work and usually result in wastage of scarce capital and manpower.

Consider the fate of the battery plant which satisfied a month's demand in 5 days. Or the $2 million date processing plant which remained unproductive for 2 years because of a blow-out in the cleaning and destaining unit supply which no one could repair. Or of the confectionery plant which lies dormant most of the year because 80 per cent of the sales are made during the Hajj — pilgrimage to Mecca — season. Or of the electronics firm whose production line never really got off the ground because of the rate of absenteeism among the key workers. Or of the woollen textile factory with 10 per cent material wastage because its management had no idea how to set and control material usage standards. Think of these and you will know why imported technology does not work in the Muslim world and why it cannot *really* work in the long run.

Passive technology transfer can only lead, on the one hand, to accumulating problems, and on the other hand, to accumulation and adjustment of metropolitan 'pushers' of technological hardware. More dangerously, forced unsuited technological change could be culturally destructive as well as lead to depletion of energy resources. Let us elaborate on this.

In Muslim societies, advanced technology, by reproducing more rapidly and more cheaply, produces more unemployment by replacing manpower with machines as well as by underselling the craftsmen who use traditional methods to produce the same goods. However, real income is reduced if the advanced machinery requires a high proportion of imported parts. But, as often the imported machinery does not function at its best, due to lack of managerial skill and educational abilities needed to operate the specialized equipment, it is run at an economic loss. The end product: increases in unemployment and decreases in real income. As an example consider the case of the country where a firm imported machinery for manufacture of plastic sandals, producing 1 to 5 million pairs a year with a labour force of 50. These slippers not only undercut the traditional leather suppliers in the domestic market, but they also forced out of employment 5,000 cobblers. This in turn reduced the market of local suppliers of leather, cotton thread, hand tools, glue, wax, polish, laces, etc. As plastic for the new industry had to be imported, there was no net income from the imported technology. So the net result was increased unemployment, wastage of scarce capital as well as considerable social disruption.[13]

When a multinational corporation transfers technology to its subsidiary in a developing country, the arrangements are usually unwritten. Joint enterprises are, however, covered by explicit agreements. Nevertheless, information on direct costs—all costs up to the full operation of the enterprise—is often inadequate and untrustworthy. UNCTAD figures show that in 1968, $1,500 million were paid out by the developing countries on direct costs. This figure has been rising annually by 20 per cent.

Indirect costs of technology transfer are even higher, and difficult to measure. They are accounted by (1) overpricing of imports; (2) the profit due to capitalization of the 'know-how'; (3) the profits that are repatriated by foreign-owned firms price 'mark-ups' for technology including the increases in the cost of imported capital goods and equipment.

Other costs of technology transfer result from (1) limitation imposed on transfer arrangements, (2) transfer of inappropriate technology, (3) delayed transfer, (4) 'non-transfer' and of course, (5) the deflecting effect on policies away from a sound development of local technological capabilities.

When all is said and done, technology transfer makes no economic sense. It is difficult for us to imagine that transfer of technology will make any poor country into a rich one. On the contrary, it leaves the developing countries at the mercy of the Occident. As everything from raw materials to spare parts to managerial skills have to be imported from the Occidental countries, the developing countries are reduced to the status of 'dependants'. Majid Rahnema sees this technological dependence as a new form of colonialism:

> . . .the present moment of history is for all of us a kind of strange purgatory. We have just emerged from an era of entire or partial dependence upon colonial powers, but our present struggle for development takes place precisely at a time when we have to rely on Western technology in order to preserve our future freedom. Such a reliance is of necessity and of our free will, yet it is also a fact that, as a result of our weakness, this might ultimately lead us to a new colonial domination of a much more subtle nature. It is, therefore, essential that we do not play a passive role in regard to technological

supremacy.[14]

The fact is that conventional occidental technologies are not designed for needs or the capabilities of the Muslim countries. They do not take into account the unique situation of the Muslim countries, the means most suitable for exploiting the particular raw materials that exist in the Muslim countries, or for the development of processing methods suitable for these particular raw materials. How, then, can we possibly expect these technologies to work for us and bring us out from the pit of underdevelopment?

There are at least two other reasons for not importing conventional technology; it is culturally subversive as well as downright destructive.

Technology and Cultural Subversion

It is commonly believed that technology is a basic necessity for 'progress' and freedom from the calamities of nature. However, technology does not only set free, it also enslaves. And the freedom it provides is not freedom from the hazards of nature; it is freedom from all transcendental values. At the same time it confines man's thought to all that is technical and mechanical. This technocratic, mechanistic and reductive outlook flourishes as a system where individuals can develop providing they surrender unconditionally to the technical organization.[15]

Once you have surrendered to technical organization you have allowed technology to be subversive. The first front is opened in the governing and administrative activities of the state. Here technology transforms the whole organization of the military and civil service. This mechanization appears to increase the power of the state; and indeed it does even to the extent that renders negligible any disadvantages that may be involved. But it is precisely this increase in the power of the state that should warn the thinking individual. It comes to the state not as a gift but as a loan from which technology expects to gain. And gain it certainly does. The end result may well resemble the *Brave New World*[16] of Huxley or the nightmare of George Orwell that is *1984*.[17]

When a piece of technology like Concorde lands in Bahrain, it brings with it the ideology of its makers. The transfer of ideology goes parallel with the transfer of technology. A transferred technology impresses itself on the cultural patterns of the host country in a way which Wells,[18] amongst others, has demonstrated. His study of technological choice carried out on a sample of fifty industrial plants in Indonesia (covering such industries as soft drinks, cigarettes, plastic sandals, and bicycle tyres), shows that often the technology selected in any one instance was *not* the most appropriate for the prevailing conditions of the economy, production and organization. Their selection was motivated by ideology more than anything else. Strip off the guise of neutrality, allow the political motivation to come to the fore, and you can see technology as a means of social control.

Today, Concorde is the symbol of speed and modernity. This mobility is a sign of progressive mass formation, which has come to mean the same thing as technological progress. But as we get mobile, we also get mobilized. Not just that, we also become mentally mobile — that is wide open to the onslaught of 'isms' and invasions of ideologies.

The impressionability of a large segment of the population to 'isms' and ideologies and the power which the demogogues derive from this is symptomatic of mass formation. 'Isms' and ideologies are generalizations, vulgarization of faith and knowledge; as such they are highly mobile, easily transferable and very infectious. As most technologists do not concern themselves with anything other than their specialization, the ideology fills a vacuum, bridges a gap. So, wherever there is occidental technology, there are occidental ideologies in the form of 'capitalism' or 'socialism', 'fascism' or 'communism'. 'Islamic Socialism' and 'Islamic democracy' are but two variants of the imported 'isms'. The relationship between technology and ideology is a powerful one. And the ideology gains most of the power which it directs towards a chosen goal.

Technology is like fire. As long as it is under your control, you can derive benefit from it. Let it get out of hand, and you will be the first one it will destroy. And then the trees, and then the woods. And finally the earth itself.

As most of the goals of contemporary ideologies and 'isms' are in conflict with Islam, the result is an acute tension with Muslim societies. Those motivated by Islam wish to go in one direction; while those motivated by technological ideologies choose another. The outcome is often predictable.

The apparently rapid advance of 'technological progress', say in the nations of the Middle East, creates an optical illusion, deceiving the observer into seeing things which are not actually there. Technology may be able to solve certain problems, declination for example, but we must not expect it to achieve that which lies beyond technical possibilities. Since even the tiniest mechanical process consumes more energy than it produces, how could the sum of all these processes create abundance? Technology produces only illusory riches. What it really does is to produce a steady exponentially growing consumption. This consumption manifests itself in desire and needs — and waste.

It creates desire for 'more' and 'better' products. The production orientation of conventional technology means in economic terms that one cannot sell the consumer goods pouring out of existing factories unless there is simultaneous desire to invest more capital and resources in new factories to make more goods. Alternatively more purchasing power must be provided to the market by inflationary spending on non-marketable products. This characteristic of not being able to use what is already there for the satisfaction of present needs unless time and resources are devoted to different future needs now pervades all aspects of technologically advanced countries. Technology has built-in feed-back loops which require investment in future technology to avoid sudden collapse of existing technologies. In other words, technology feeds on itself. Eventually it becomes a 'superstar'[19] and takes its own course.

This runaway technology creates waste. Waste is the most blatant by-product of the production-orientated technology. The continuing desire for more and more is, of course, the other side of the equation. The more you produce, the more you waste. And where there is waste, there is desolation. Technology invades the landscape with destruction and transformation: it grows factories overnight, builds pre-fabricated cities, cities indescribably hideous and ugly, where the misery of man is plain for all to see. Cairo, Karachi, Algeria and Ankara are reflections of New York, Glasgow, Tokyo and Turin. Here technology has polluted the air, poisoned the water, wiped out plants and animals. The scales may be different, but the effect is not. Urbanization is equally fatal everywhere. Soon in the cities of the Muslim world, too, nature will have to be 'protected' and 'preserved' from the effects of technology. Large areas of green will be set apart and fenced, placed under taboo, like museum pieces. The very fact that areas have to be protected implies that a destructive process is at work.

And how can we meet the goals of the economic principles of Islam — of reducing consumption, cutting down needs, of reducing savings and investments, of spending more in the way of Allah — with conventional technology at the very foundations of the strategies for development?

Technology Assessment

We believe strongly that the economic principles of Islam constitute a case for an entirely different pattern for technological activity. We do not reject technology *per se* for technology is needed even to produce safety pins. But we do reject occidental technology, and we should. We shall return to the characteristic of the new technology later.

To be realistic, the Muslim world's dependence on conventional technology cannot be reduced overnight. It can only be reduced, and eventually eliminated, with time while indigenous technological capabilities are developed.

Of course, the characteristics of a new technology which replace the conventional occidental technology in the Muslim world, can only be derived from the strategy of development we outlined in Chapter 3. There we stated that all development activity must be subject to the principle of domesticity: the real best interest of a culture, the preservation of a society's cultural integrity, should be the guiding light of the governing and scientific communities in a country. When it becomes absolutely necessary to import technology, we suggest that the Muslim people treat the import with the same discrimination they eat 'glab', the water melon seeds, so popular in the Arab world: that is, split them with the teeth, swallow the kernel, and spit out the shell! In other words, strip the import of its values and cultural bias and accept only the hardware that is suited to the local condition. More specifically, the principle of domesticity requires:

(1) an awareness of the value-systems inherent in the technologies;
(2) self-confidence in the cultural sufficiency of the society;
(3) agreement that undesirable changes in the society shall not take place;
(4) selection and scrutiny of imports;
(5) their modification, if necessary, before adoption;
(6) the monitoring of the impact of the imports; and
(7) their rejection and abandonment if they prove corrosive.

To fulfil these criteria, it is necessary to have some idea about the long and the short-term, planned and unforeseen, direct and indirect, social, ecological, economic and political costs and benefits which may accrue to a society if it adopts or purchases a particular piece of technology, before we can make any decision upon it. Some of these questions will be:

1. Where does the technology currently stand?
2. In what direction it seems that it is developing?
3. In what ways is the technology likely to be applied?
4. What factors are likely to influence that application?
5. What are the likely secondary and tertiary consequences of the application?
6. What benefits could be derived and what cost would be incurred if indigenous efforts were to be made to alter either the technology applications or their subsequent consequence?
7. Which community or interest groups are most likely to be affected by the anticipated impacts of the technological applications and are most likely to take action to influence these impacts?

Questions such as these have to be answered if we are to have some idea of the effect on Muslims as individuals and on their total environment of a particular candidate for import. Needless to say there is no accepted method for conducting the multidimensional type of analysis that is needed here. However, it is not necessary to utilize all the latest sophisticated computerized techniques to develop a capacity for this type of technology assessment. Methodologies can be developed on a human scale to meet the needs of specific countries.

The techniques currently applied for technology assessment in the Occident are all based on the neo-Apollonian outlook we discussed in Chapter 2. In this type of assessment all decisions must be justified in terms of 'rational factors' and 'valid theories' (of occidental science and social science). Such methodologies incorporate occidental value judgements and only obscure the fact that the categories selected as the basis for analysis already disguise the essential ideological and political elements of the problem under review.[20] The analysis in question would be culturally biased *against* the Muslim societies.

Basically, technology assessment is a policy tool; and like all policy tools there are several assumptions at the conceptual foundation of technology assessment. These suggest that:

(1) Value judgements are made in the collection of data, as well as its evaluation and analysis;

(2) for effective policy formulation it is necessary to organize certainty and uncertainty in order to define effective strategies for managing any particular technology;

(3) more information and analysis promotes better decision-making; and

(4) in the long run, indirect and unanticipated effects of technology are often more significant than the immediate planned
consequences.

With an awareness of these assumptions, the Muslim countries can
develop the modalities through which they can apply this evaluative
tool in their own planning and policy-making. Here we would like to
emphasize that assessment should highlight cultural bias, uncertainty,
ignorance, and risk. Decision-makers will have to make a choice on the
basis of the principle of domesticity.

Developing Indigenous Technology

Let us now return to the new technologies that would meet the goals
of the economic principles of Islam. We shall call these new technologies 'indigenous technologies' to reflect their local character. Their
'newness' will become apparent as we proceed with our discussion.
Some major characteristics of indigenous technologies will be:

(1) they will be capital-saving and labour-intensive;
(2) they will be cottage-scale and small-scale, taking account of
the human factor;
(3) they will be directed towards producing goods and services
appropriate for large-scale consumption rather than for individual
luxuries;
(4) on the whole they would be based on simple processes; where
necessary these processes will be adequate modifications of traditional skills like pottery, weaving, etc.;
(5) they will rely on the use of local material, rather than imported materials;
(6) they would be energy saving rather than energy intensive;
(7) they would utilize locally available energy resources such as
sun, wind, water and bio-gas;
(8) they would promote a symbiotic and mutually reinforcing,
rather than parasitic and destructive, dependence of metropolitan
city upon the rural population;
(9) and finally they would be non-violent, based on rational
sustained use, rather than indiscriminate rapid devastation, of the
environment.

We noted in Chapter 7 that small, labour-intensive firms are most
suited for Muslim countries. If the technology employed in these firms

is based on simple processes, within the grasp of the local population, and if the manager is also the owner, the need for complex managerial skill becomes redundant. Also the need for foreign employees, draining away vital foreign currency reserves, evaporates. Furthermore, as small firms are less dependent on large towns, they can be located in rural areas producing even development throughout the country.

Indigenous technologies could be very 'progressive' if they could be reproduced by local industries in a short span of time and if they used indigenous raw material. This would be relatively easy if indigenous research capabilities were fully developed. This is, of course, also an argument for further developing local research capabilities and reducing the utter reliance on imported technologies. Developing local capabilities would also provide opportunity for talented native individuals to expand their potential. In the long run new indigenous technologies should create local industries for machine tools, machine building and repair, and material and component manufacture, thus building a strong base for indigenous technologies to flower further while at the same time keeping within the capabilities of the local population.

Since Muslim countries, by and large, have either a large amount of unemployed or underemployed labour as well as a capital shortage, indigenous technologies should place emphasis on maximizing capital productivity. The method chosen for making a given product should provide the highest output for a given capital cost.

Ademola Banjo lists the conventional arguments against small-scale, labour-intensive, capital saving technologies, as follows:

1. If people have to learn new technologies, they might just as well learn the most modern and effective ones, as the intellectual efforts are essentially the same.

2. The cost of a particular plant installation with alternative technologies may be low, but because of lower productivity the actual capital cost per unit produced would often be higher. Hence the final cost per unit produced will also be higher.

3. The maintenance cost of secondhand and obsolete equipment is higher and its operation often needs a higher level of individual skill. Furthermore, spare parts could become unobtainable and would have to be especially made at greater cost.

4. Only by raising productivity can employment be increased. This would bring cost and prices down and hence promote consumption and the ability to export in competition with similar products.

5. Development implies social change and dislocations are part of

the process of social change. These can be taken care of by appropriate social policies and programmes. In any case, many communities are prepared to pay the price of social disruption, if necessary, to raise their standards of living as soon as possible closer to those they already perceive through the modern communication media.[21]

This type of criticism overlooks the fundamental basis of indigenous technologies, that is, to replace production-orientated, capital-intensive, technologies with capital saving, labour-intensive ones. Furthermore, indigenous technologies, even the most simple ones, will be derived from local research and development. Simple technologies do not always imply obsolete imported technologies. As E. F. Schumacher has pointed out, anyone can complicate things—it takes a touch of genius to keep them simple.[22]

The spread of indigenous technologies in the Muslim countries is thwarted by four factors. Firstly, as we pointed out earlier, there is too much uncritical acceptance of capital-intensive methods. The will to systematically explore alternative possibilities seems to us to be conspicuously lacking in the Muslim world. Secondly, the trade and commercial policies of the Occident have an adverse effect on the development and spread of labour-intensive technologies. The developed countries have all but closed their domestic markets to the products of labour-intensive technologies. This makes economic co-operation between Muslim countries even more desirable. Thirdly, public policies pursued in the Muslim countries encourage the use of capital-intensive technologies. The maintenance of highly overvalued exchange rates and a monetary policy of keeping rates of interest substantially below the level of that in the unorganized sector of the money market, are two policy measures widely used in the developing countries in general, and Muslim countries in particular. This encourages business enterprises to use technologies and processes that are more capital-intensive than they would if exchange values and interest rates were at their proper market levels. This bias is further aggravated by the high salaries that occidentally trained management, as well as organized labour, often manages to obtain for itself. Fourthly, in many countries the most powerful force holding up labour-intensive technologies is the occidentalized ruling elite. When there is polarization between the ruling elite and foreign business interests on one hand, and traditional intellectuals on the other, no motivation can then exist for introducing improved technologies.

In view of the four negative factors described above, what specific measures can be taken to create a conducive environment for the development of indigenous technologies? Furthermore, in which sectors can the philosophy of indigenous technologies be made operational immediately?

There is a whole range of human activity where the outlook of the new technologies can be introduced without any real effort. To take the seemingly simple problem of building construction: is it really such a formidable problem to design structures and processes which utilize local materials, traditional skills and produce small human-scale constructions in harmony with traditional backgrounds as well as the environment? Is it really difficult to apply mechanical engineering design to the production of small, easy to build, operational tools and systems for the improvements of agricultural tools, and operate rural transport facilities? Amongst a multitude of projects that can be undertaken to develop new technologies, indigenous capabilities and local self-sufficiency could be included the following:

(1)   application of civil engineering to the rural environment—for example, development of low-cost building technology based on local materials;

(2)   development and exploitation of locally available energy sources, such as the wind, the sun and bio-gas, and the rational management of fuel sources such as forests;

(3)   application of mechanical engineering design for the improvement of agricultural implements and operations, and of rural transport facilities;

(4)   development of small-scale industries to exploit local raw materials and agricultural products and wastes;

(5)   development of small-scale industries to produce consumer goods for urban markets;

(6)   development of techniques, books and low-cost learning materials for education in rural schools; and

(7)   application of the environment and ecological approach to rural resources such as soil, water, grassland, forests, livestock, wild life and fish.[23]

Following K. Marsden in part,[24] we list some possible policy measures which could ensure that indigenous technologies are utilized efficiently.

1. Providing indigenous industries with a scope to expand, develop and diversify over time without clashing with competing industries which are technically more advanced because greater

resources (uneconomically priced) have been placed at their disposal. Giving a clear run ahead to indigenous entrepreneurs is likely to be more conducive to growth and development than disposing protective subsidies and quotas in an attempt to have the best of both worlds.

2.   The setting up of documentation and information centres to keep track of past and current technological developments throughout the developing world in general, and the Muslim world in particular. These would establish close liaison with international and other national advisory services for the selection of equipment.

3.   The provision of widespread primary and technical education facilities at the apprentice level, combined with night school tuition and upgrading courses for practising operatives, supervisors and managers.

4.   The encouragement, by state subsidies, grants, etc. of trade and research associations for each industry, sponsored and run by the members themselves. Special budgets could be allocated for importing standard machines to be stripped down, adapted and eventually reproduced locally.

5.   The institution of incentive rewards schemes for inventions, as well as patent protection for local adaptations of foreign designs.

6.   The formation of common facility co-operatives and joint production workshops to raise the productivity of artisan and handicraft industries.

7.   The provision of extension services for small-scale entrepreneurs, providing advice on product and process development, technical skill formation and the selection and use of appropriate technologies.

8.   Long-term planning of manpower and skill requirements in the various sectors of the economy, closely related to the foreseen rate and character of technical change.

9.   The adoption of factory legislative and safety regulations which provide adequate working conditions and safeguards for all groups of workers but do not create dual standards (i.e. for those within and those outside the practical jurisdiction of the laws) and act as barriers to expansion for the smaller enterprises.

10.   The creation of central quality control and inspection schemes to ensure that products destined for export meet external quality standards, but without imposing unrealistically high standards on total production within the country.

11.   Priority in the allocation of import licences for machinery

and materials to organizations that have already demonstrated the aptitudes, skills and motivations required for success in the export markets.

12. The establishment of special small business Islamic development banks to reduce the differentials in capital accessibility between the traditional and the modern sectors.

13. The planned distribution of industry of 'backward areas' to provide more employment opportunities outside the major cities and to reduce income inequalities between regions. Processing of agricultural and other land-based products are obvious choices.

14. Subsidized factory premises in provincial towns and villages to slow down the population drift to the cities. The subsidies could be equivalent to the cost of housing and other facilities which otherwise have to be provided in the cities.

15. Public information campaigns to increase the prestige and consumer acceptance of indigenous technologies and products.

It seems to us that, for the Muslim countries, there is only one choice: to undertake those activities which ensure the flowering of innovation and development of indigenous technological capabilities rather than adjustments to imported technologies. However, just as important as development of local technical skills is the need for imagination, intellectual boldness, experimentation and theory building. These aspects of our culture, which must stamp our technological capabilities, are the basic guarantees for successful development and our cultural survival.

Notes

1. H.W. Fowler and P. G. Fowler (eds.), *The Concise Oxford Dictionary*, (Clarendon Press, 1964), p. 1330.
2. E. G. Mesthene, *Technological Change* (Mentor, New York, 1970), p. 25.
3. J. K. Galbraith, *The New Industrial State*, (Penguin, 1967), quoted by D. Dickson, *Alternative Technology and the Politics of Technical Change* (Fontana, London, 1974), p. 205.
4. Edward de Bono, *Technology Today* (Routledge and Kegan Paul, London, 1971).
5. Ibid., p.205.
6. Ziauddin Sardar and Dawud Rosser-Owen, 'Science Policy and Developing Countries', in I. Spiegal-Rosing and D. de Solla-Price, *Science, Technology and Society* (Sage Publications, London, 1977).
7. From M. Okano, *Les Nouvelles*, June 1972, p. 27
8. Leslie Sklair, *Organised Knowledge* (Paladin, London, 1973), p. 260.
9. Dickson, *Alternative Technology*.
10. As an example, see A. C. Sutton, *Western Technology and the Soviet Economic Development 1930-1945* (Hoover Institution, Stanford, 1932).

11. See Lynn White, 'Historical Roots of our Ecological Crisis', *Science* 155, 1203 (10 March 1967), and E. Mendelsohn, 'Should Science Survive its Success?' in R. S. Cohen *et al.* (eds.), *For Dirk Struick*, D. Reidel, pp. 373-89.

12. G. K. Boon, 'Choice of Industrial Technology. The Case of Woodworking', in *Industrialization and Productivity* (UN, 1961).

13. K. Marsden, 'Progressive Technologies for Developing Countries', in W. Galenson (ed.), *Essays on Employment* (ILO, Geneva, 1971).

14. Majid Rahnema, 'Iran: Science Policy for Development', *Impact of Science on Society*, 19, 53-61 (1969).

15. See, for example, John Wilkinson (ed.), *Technology and Human Values* (Centre for the Study of Democratic Institutions, Santa Barbara); Ivan Illich, *Tools for Conviviality* and *Celebrations of Awareness* (Calder and Boyers, London, 1973 and 1971); T. Roszak, 'Science — A Technocratic Trap', *Atlantic* 230 (1) 56 (July 1972): J. H. Muller, *The Children of Frankenstein* (Indiana University Press, Bloomington, 1970); John G. Burke (ed.), *The New Technology and Human. Values* (Wadsworth, Belmont, 1966).

16. Aldous Huxley, *Brave New World* (Chatto and Windus, London, 1932) and *Brave New World Revisited* (Harper & Row, New York, 1965).

17. George Orwell, 1984 (Secker and Warburg, London, 1949).

18. L. J. Well, 'Economic Man and Engineering Man: Choice of Technology in Low Wage Country', *Economic Development Report*, Autumn 1972.

19. Council for Science and Society, *Superstar Technologies* (Barry Rose, London, 1976).

20. This is true of methodologies described in such works as François Hertman, *Society and Assessment of Technology* (OECD, Paris, 1973); and Joseph Coats, 'Technology Assessment: The Benefits, The Costs, The Consequences', *Futurist* 5 (6) (December 1971).

21. Andemda Banjo, 'Pros and Cons of Intermediate Technology', Technology and Development, *Unitor News* 6 (4), 11-12 (1974).

22. E. F. Schumacher, 'Economies should Begin with People, Not With Goods', *Futurist* 8 (6) (December 1974).

23. Cf. A. K. N. Reddy, 'Alternative Technology: A Viewpoint from India', *Social Studies of Science* 5, 331-42 (1975).

24. Marsden, 'Progressive Technologies for Developing Countries'.

9 R & D: BASIC OR APPLIED?

The realization that the problem of development in the Muslim countries is coupled to local flowering of innovation and technological capabilities is very recent indeed. Many Muslim countries, as a result of this realization, have either established, or are in the process of creating, institutions which encourage development of local technological capabilities and help in the integration of science and technology programmes with local needs and capabilities. Iran, Egypt and Saudi Arabia, for example, have established ministries responsible for science. Algeria and Tunisia have created departments responsible for science in established ministries. In Turkey, the State Planning Organization, in Indonesia the Indonesian Institute of Science and in Nigeria the Nigerian Council for Science and Industrial Research have been especially created to deal with the area of science policy. These agencies and institutions as well as the governments of Muslim countries will face problems of choices about the allocation of resources for Research and Development (R & D). There are no clear-cut rules for such choices; neither are the examples of the nations of the Occident particularly helpful to the Muslim world. In this situation, what criteria can the Muslim countries use for allocation of resources?

What Is Pure? And What Is Applied?

The *modus operandi* of the system of research and development is not clearly understood. It is not easy to evaluate the inputs into, and more particularly, the outputs from the system.[1] However, it is generally believed that the connection between research and its subsequent application is almost never direct but almost always indirect. The application of research is brought about by such factors as institutional overlaps between science and technology and by entrepreneurial attitudes towards newly acquired knowledge. Ideas are supposed to emanate from the 'pure' or 'basic' end of the R & D spectrum, and terminate at the application end.

In this spectrum, 'pure', 'basic' and 'fundamental' research exists at one end and 'applied', 'mission-orientated' research at the other. There is no set of rigorous operational definitions for what scientists

139

are actually doing when they are supposedly engaged in either kind of activity. However, pure research comes in the category of science for the sake of science. This is research on the problems that are considered 'intrinsically' interesting and where the utilitarian aspects play only a secondary role. Applied research, on the other hand, aims at satisfying certain needs or solving certain problems. For this reason applied research can be assigned to a certain specific area, or even to a handful of areas. The decision whether or not to undertake applied research of a certain type depends on the perceived benefits it can bring.[2] Roderick defines

> ...fundamental (basic) research as the study of nature in order to understand it, and applied research as the study of nature in order to control it. It is obvious that before you can control something you must understand it, but it is often possible to partially understand something and be able to control it in a reasonable manner.[3]

Thus, pure science is taken to be universal, value-free, international in outlook. It is universal because it has been discussed according to strictly objective rules and procedures of proof and reproducibility. The entire community of science participates in acknowledging this discovery.

With this definition of pure science the neo-Apollonians often argue that developing countries should not waste their resources in basic research. They should freely borrow what they need. Thus Harvey Brooks:

> Basic (pure) science, unlike technology, is truly international in its value. 'Know-how' does not have to be purchased but freely shared among all who are intellectually prepared to appreciate and use it. In consequence the basic research in one country can be readily appropriated for application in another...[4]

There are several reasons why such advice is highly dangerous. Research in the industrialized nations is not carried out randomly: pure or applied it is geared to the needs of the Occident and the emphasis bears a relationship to its potential value to the funding country(ies). We have already argued (in Chapter 2) that science is culturally biased and, as such, to assume that the fruits of occidental research can be easily applied to developing countries is to be a dupe of somnambulant wishfulness. This assumption not only creates an illusion but also isolates one component of indigenous science research from another. In fact only 2 per cent of all the scientific research that is carried out in the Occident has any relevance to the developing countries.[5] How much of this would have any relevance for the needs of the Muslim world?

Even that minute segment of occidental scientific research which may be useful to Muslim countries may not be actively applicable due to local conditions. As an example: high yield varieties of corn or rice may not thrive in temperate soil and climatic conditions. Successful production would require further research on local factors.

In the Muslim world today, science is sporadic, isolated, largely unconnected with local needs and interest and quite incapable of self-sustenance.[6] This is often attributed to bad planning, faulty administration by government agencies and misplaced priorities.[7] This is partly true. Partly it is due to the lack of funds. Studies in R & D expenditures in occidental countries have shown that in order to conduct a 'modest' programme of research in the 1960s a country had to spend between 0.7 and 3.5 per cent of its GNP on R & D and to have 4,000˙ scientists and engineers working on R & D per million inhabitants.[8] In contrast, the Muslim countries have, up to now, spent less than 0.5 per cent of their GNP on R & D and have less than 10 scientists and engineers per million inhabitants. This is only the norm. There are deviations from this norm: Nigeria, for example, had in 1972 R & D expenditure of 1 per cent, that is US $50 million; Turkey, in contrast, spent 0.35 per cent of its GNP on R & D in 1972.[9]

However, a more important reason, in our opinion, for the fragmentation of science in Muslim countries is the result of the inappropriate attitudes on the part of the local scientists who have taken to heart the advice of Harvey Brooks and colleagues. They either sit idle waiting for their 'fair share' of the 'know-how' or pursue research which identifies with the 'high culture' of 'international science' but has little or no relevance to the local needs and problems. The few pure research centres of quality that do exist in the Muslim world function as isolated enclaves, doing little for local R & D. The research staff have little or no motivation or incentive and by and large they continue the work they did during their training in the Occident.

One outcome of this situation is that the local industry does not have confidence in the results which these centres can produce while the research centres themselves lack awareness of the needs of their potential clients.

An equally tragic outcome of this isolation and fragmentation of science in the Muslim world is the well-known 'brain drain'. Since science is not 'consumed' at home, the only 'satisfying career' for the up and coming scientists is found in the corridors of occidental academies or industry.

We believe R & D in science to be very much a complete system: various components of this system, pure as well as applied research, local technological capabilities and industrial innovation, must combine to evolve a coordinated planned structure. It is the active lack of this integration of the various parts of local science efforts of the Muslim countries in a *coordinated operational system* that prevents their science from developing and being productive. Furthermore, recommendations from the Occident about what a Muslim country should do in the fields of R & D often lead to the gulf between the component parts becoming much wider.

The interdependence between science and its application does not occur automatically; it must be planned, nourished, and developed. Science must be made to serve indigenous goals; and because the resources are limited — either in finance, or in manpower, or both — areas of research must be studied and planned. In this respect both basic as well as applied research are important for the Muslim *Ummah*.

For the purpose of this chapter, we shall divide basic research into two categories: target-orientated basic research and conspicuous basic research.

Target-orientated pure research is used here to refer to those elements of fundamental research which have, or may be expected to have (or ought to have) direct or indirect importance for the Muslim countries. In other words, it is pure research into the specific present and future needs of the Muslim *Ummah*. Such research needs are normally detected in the course of practical planning: the term 'target-orientated' focuses on indigenous research priorities which are singled out from the needs of the community as well as during planning.

Conspicuous basic research is quite similar to conspicuous technology: it has high prestige value but little significance to local needs. Those scientists in the Muslim world who engage in international research and debates on theoretical and particle physics or seek the great breakthrough in cell protein or cancer chemotherapy are all involved in conspicuous pure research. Unfortunately, there are just too many of them.

The most distinguished scientists from India and Pakistan found at international symposia and conferences generally represent the advanced fields of research and not the less fashionable subjects of agricultural research or fisheries or other such fields so close to the heart of an economist and so often forgotten by the scientist.[10]

The politics of science dictates that science is funded so that it may benefit the nations. The policy-makers rightly believe that science will solve local problems of production, water supply, malnutrition etc. As a result it is natural that political and economic criteria are used in setting priorities in national science policies. In the Occident, large-scale basic research is funded either on the assumption that it will eventually 'feed' a large part of technology or on the dictates of fashion. In Muslim countries research cannot be funded on the basis of hopes or fashions. It must be target-orientated: it must endure a system of research which takes account of indigenous needs, requirements and interests as well as promotes the application of research in local industry; it must be related to problems of national urgency. This, let us hasten to add, is not a question of political interference with the contents of science, but about the use of science as one of the principal instruments of national development.

When examined from this viewpoint the whole argument about basic and applied science appears irrelevant and misleading. As Nancy Stepan points out:

> . . .the history of sciences in Europe and the United States shows that in the last sixty years or so distinctions between basic (pure) and applied science have been breaking down and the frontiers between them disappearing. While there are some research investigations that are more general and theoretical in their conclusions than others, and other investigations that have very specific applications as their goals, the most important aspect of science in the industrial world is that science and its applications are ultimately intertwined. The decision to call a piece of work basic rather than applied often depends more on the type of scientist carrying out the work and his place of research (university or industrial laboratory) than the character of the work itself.[11]

In some fields such as biomedicine, geo-physics, agriculture, fisheries and soil mechanics distinctions between basic and applied research seems particularly inappropriate. This being so, the arguments in favour of creating indigenous institutions in Muslim countries in such fields as biomedical, petroleum and agricultural sciences are especially strong.

And while few Muslim countries can afford to support expensive, large-scale conspicuous pure research, it is evident that attempts must be made at home to create an interlocking system of basic and applied research in accordance with the principle of domesticity. In this respect, the rules suggested by Michael J. Moravcsik for developing indigenous research capabilities are worth mentioning:

1. Local advantages should be utilized. These may be geographical, climatological or astronomical, based on some easily available raw material, etc.

2. Intrinsic costs should be compared; some areas of science are inherently more expensive than others.

3. Potentials for application should be compared; some sciences, viewed from our perspective, are more remote from short-term application than others. However, such judgements are unreliable. Nuclear physics, extremely esoteric in the 1930s, suddenly produced one of the main technologies in the 1940s and 1950s.

4. Educational objectives should be kept in mind. Some areas of science are better suited for educating students than others, where, for example, specialization is too narrow.

5. Attention should be given to the general considerations for determining scientific choices in terms of the extrinsic and intrinsic potential of the subdisciplines. These considerations also apply to science in LDCs, though they should not be the exclusive determinants in the selection of research areas.

These guidelines [and others mentioned in the second section of this chapter] can be used in making a list of preferred areas of research. There is, however, one consideration with which all these criteria pale in comparison — the availability of outstanding people. Scientific activity centres around creative individuals. Such individuals in the existing scientific manpower of an LDC are an asset that must be utilized. At the beginning, scientific activity is weak, critical mass is questionable, and the indigenous scientific community consists mainly of inexperienced researchers with a lack of leadership and direction. The presence of an outstanding person, able to serve as a focal point for exciting scientific activity, is invaluable and should largely influence the areas of research supported in that country. Such persons will generally have definite ideas of what scientific problems they consider interesting and should be supported in research on those problems unless it is financially impossible.[12]

Quantity *v.* Quality

Basic or applied, the most acute problem associated with research in the Muslim countries is manpower. In Indonesia, for example, the output of science graduates from 24 out of 40 state universities between 1950 and 1967 was 11,050 in 'exact' science, 9,345 in 'non-exact' sciences, and 1,417 in teaching sciences—a negligible output for a country of 100 million inhabitants.[13] In Turkey, the total numbers of engineers, architects and agricultural professionals were 6,200 in 1950, 10,000 in 1958, and 15,460 in 1969.[14] Clearly, when faced with such abysmal statistics, it is essential to make an all-out effort to increase scientific manpower.

Planning for manpower is, of course, a long-range exercise. However, the desire of many Muslim countries to leap-frog to the occidental living standards often persuades them to take short-term measures rather than plan constructively for long-range development.

In the short run, it is indeed possible, provided that financial backing is available, to set up 'overnight' an impressive array of research institutes, power plants and large factories.[15] However, it is another matter to produce a core of local personnel to man them. Successful 'production' of sophisticated manpower is a multigenerational process requiring practical planning over several decades.

Just as important as quantity is quality of manpower produced. For over two decades, the scientific establishment of the Muslim world has been described by only one word: mediocre. Michael J. Moravcsik and John Ziman find 'no more than the fragments of a scientific community, disorganised, disunited, of limited professional competence, poverty stricken, intellectually isolated, and directed towards largely romantic goals—or no goals at all'.[16]

The present establishment consists of either overtly narrow specialized scientists who trained in the Occident or home-grown scientists who have come to be dependent more on memory than original thinking and who react negatively to scientific rigour and even traumatically to any suggestion of practical or applied work. The lack of quality leadership perpetuates inappropriate acceptance of research priorities of the industrial countries. The reliance on the Occident on problem definitions, much less research plans, means that key areas of (1) education and training, (2) identifying and defining key problems, (3) adopting technology to their own special local situation, and (4) increasing communication between the scientific community itself as well as between scientific community and government officials, are never developed.

What is more, this mediocre establishment of science has a stubborn quality of self-perpetuation. Those with dubious standards are not very keen to allow promising new potential to enter their research centres. University teachers and research supervisors breed poor successors. Governments place false pride in these mediocre communities: impressive population statistics are produced to justify the rapid growth of the establishment of mediocre science. Such arguments, naturally, miss the point that too rapid growth must increase the proportion of mediocre scientists.

Science development is a slow and painstaking process. It requires a delicate balance between quality and quantity. As such, we believe that the rate of growth of science in the Muslim world should be moderate. Every step is to be planned and planned in some detail and with some care.

Scientific research does not take place in isolation. It requires an adequate physical environment including experimental equipment, an auxiliary support system—workshops and technicians, libraries and information units, computers and programmers—and channels for communication between scientists themselves, between scientists and governments and between scientists and the public. The quality of the end product, to a certain extent, is dependent on the quality of the support system.

The Muslim world, up to now, has paid little attention to the support system of science research. This is quite evident from a casual perusal of their development plans.

The lack of this support system for research is an important contribution towards the brain drain. It is commonly believed that a large number of good scientists who leave their native lands do so for financial reasons. This is certainly true for some. However, as the majority of Muslim scientists, particularly those trained in the Occident, enjoy high salaries and a much higher standard of living than their average countrymen, this motivation cannot be attributed to all. We believe that the brain drain is caused by the lack of the support system for science research.

At this point let us also mention two other overriding factors leading to the brain drain: namely, infighting and conflict with bureaucracy. Anyone who has worked in the Muslim world knows that it is difficult to get anything done in many Muslim countries without either straightforward bribery or fully-fledged conflict with the bureaucracy. The endless battles with corruption, arguments with apathetic adversaries, and incomprehensible red tape all contribute towards persuading many

Muslim scientists that they would really be better off elsewhere. A reflection of the external bureaucratic conflict is the infighting amongst the scientists themselves. The internal battles are particularly acute between the home-grown products and those scientists who trained in the Occident. In this environment nothing can flourish; and least of all science.

To return to the lack of support system for science research in Muslim countries, we believe that here is another argument in favour of target-orientated pure research in the Muslim world. Research tailored to the resources and needs of Muslim societies in various stages of development would reduce its heavy dependence on sophisticated support systems.

However, there is no substitute for one particular support system for science research: efficient channels for scientific communications. Although there is as yet no satisfying method for measuring scientific productivity and its contribution to the development process — notwithstanding the rather naive practice of counting articles and citations as a function of scientific productivity[17] — it can be safely said that scientific communication plays an important part in scientific development.

Scientific Communication

In the universities and research institutes of the Occident, 'invisible colleges' are accepted as a crucial mode of scientific research.[18] Invisible colleges are collections of scientists, at various geographical locations and with different levels and types of training, all having common interests, collaborating, helping and influencing the direction of research in a particular area. These 'colleges' have a very informal structure but are usually found around a nucleus of well-known scientists who influence the group. The essential point is that invisible colleges act rather like a telephone receptionist connecting individual scientists with each other, helping the exchange of ideas and information. In such a situation, research thrives; in isolation research scientists often resign themselves to mediocrity. This is what actually happens to scientists in many Muslim countries. The case of Professor Abdus Salam, Director of the International Institute of Theoretical Physics, is typical:

I felt terribly isolated. If at that point someone had said to me, 'If we shall give you the opportunity every year to travel to an active centre in Europe or the United States for three months of your vacation to work with your peers, would you then be willing to stay the remaining nine months at Lahore?' my answer would have been yes. But no one made the offer.[19]

Isolation and lack of contact with other scientists is one of the most detrimental factors in the flowering of science in the Muslim countries. But even more important than lack of contact with the international science scene is the isolation from local and regional scientific happenings. Internal scientific communication within regional groups is vital for the very survival of the indigenous scientific community. It has been estimated that to be efficient and productive, a scientist must be in contact with at least three other scientists at the same level and within the same discipline. Contact within this 'critical mass' must be frequent with exchange of results and ideas.[20]

One of the most common tools for scientific communication is the journal. A predominant source of scientific and technical information, scientific journals are usually the first to carry results of current research. Since the most recent information is preferred to older, it is necessary to have access to current issues of appropriate journals. However, most widely known scientific journals are aimed at the occidental countries, where they have traditionally found their readership. Scientists in the developing countries have difficulty in publishing in these journals. There are two main reasons for this: (1) in most journals it is necessary to pay, in hard currencies, for publication; and (2) most journals specialize in fields which, with few exceptions, are outside the domains of scientists in the developing countries. Furthermore, subscriptions to some of the journals are prohibitively expensive.

In such circumstances there is a strong case for Muslim countries publishing journals for local use as well as for the use of the scientific communities of the Muslim world. We consider this to be essential despite the rhetoric of 'information explosion' and 'over-loaded retrieval channels' simply because journals are one of the cheapest methods of communication and are necessary for raising national scientific morale.[21] Moreover, the number of scientific journals may go on increasing exponentially but as long as they do not serve the requirements of the Muslim countries, there is a need for publication of science journals in the Muslim countries. And, from the viewpoint of the Muslim scientist, what harm is there is publishing a paper in an international journal as well as a local one?

In most Muslim countries library science is not well established, libraries themselves being very scarce.[22] This is particularly ironic when we consider that in the early period of Islam, almost every important mosque had its own library with stocks of theological as well as philosophical and scientific works. Arabic chronicles give us a great deal of information about these libraries and other information centres that provided back-up services for various seats of learning. We think that a considerable amount of effort and finance should be spent in the Muslim world to redevelop this lost heritage. Communication between existing libraries must be increased and there should be more coordination of acquisition policies between neighbouring libraries.

Some Muslim countries, however, have set up the bare bones of information and documentation centres. Perhaps the most well-known of these is PANSDOC—Pakistan National Scientific and Technical Documentation Centre—which was established in 1957 with the technical and financial assistance of UNESCO. Now renamed Pakistan Science and Technology Information Centre (PASTIC), it provides the bare rudiments of an information service but nevertheless manages to keep a reasonable number of Pakistani scientists aware of current scientific literature through its document procurement, translation and bibliographic services.[23]

Turkey has an equivalent organization, TURDOK which was founded in 1967 and operates under the Scientific and Technical Research Council of Turkey.[24] Other national documentation centres worth mentioning are those of Nigeria and Indonesia. The latter has been particularly enterprising in establishing two information centres in 1972: the National Scientific Documentation Centre (PDIN) in Jakarta and Bibliotheca Bogoriensis (BB) in Bogor. The centre in Jakarta coordinates the entire network and covers science and technology in general while Bibliotheca Bogoriensis is responsible only for biology and agriculture.[25]

Needless to say, all this is only a scratch on a rather large cube of steel. Similar documentation centres are needed in other Muslim countries as well as many smaller regional information units.

Along with scientific journals, professional and learned societies are also essential for the healthy development of science. Unfortunately, these too are rare in the Muslim countries and where they do exist their existence becomes an end rather than a means. Their activities are confined to formalities, back-slapping and celebrations. Often they are little more than elitist clubs. There is a need to reorganize them so that membership becomes more of a challenge than an honour, with rewards for functional tasks. It is also necessary for them to become more vocal about their professional needs and in their criticism of the scientific communities of the Occident.

Institutional Modes of Research

In the industrialized Occident, R & D work is carried out by a wide variety of institutions: universities, research institutions, industrial and government laboratories, industrial bodies, professional societies, funding agencies and equipment manufacturers. In contrast, the Muslim world lacks not only in variety but also in number.

Even in countries with relatively well-developed scientific communities, there are common complaints that those institutions and organizations perform very poorly in respect of indigenous technological capabilities. In general, R & D organizations in the Muslim countries have the following characteristics:

(1) practically all R & D is funded by the government;
(2) most research, pure as well as applied, is conducted in government and university laboratories; and
(3) the potential users of R & D, namely industrial firms and organizations, perform little or no R & D themselves.

Traditionally, research in the Muslim world has been carried out in the universities. This habitat for research was arrived at by observing the historical connection between universities and science in the Occident, and also because one finds it difficult to cite examples of good science in the industrialized nations without a university system. However, the structure of many universities in the Muslim world is such that it hinders the integration of research and its application. Many universities in the Muslim countries are much older than those in the Occident. Some of them were founded hundreds of years before a single university opened in Europe. Yet the traditions of society and industry-orientated research have been lost. Today, universities — with few exceptions (see Chapter 10) — are simply a type of education machine: churning out spoon-fed students, very poor carbon-copies of their occidental counterparts. It is difficult to see how any type of research can flourish in such institutions.

Recently, a number of attempts have been made to link universities with industry. One example of this linkage is the university-based multi-purpose applied research institutes which have flourished particularly in the Middle East Technical University in Turkey. In this recently established university, applied research is being done under contract with various industries and government enterprises. Faculty members are encouraged to conduct a wide variety of applied research projects for extra stipends. Research programmes include pilot plant demonstration units which bring the university in direct contact with industry. Although these programmes began on an individual basis, in 1972 the university created an Applied Research Centre as a focus for industry-orientated research. As the projects are self-supporting no additional cost is incurred by the university.

One variant on this type of research institute is the Technology Consultancy Centre at the University of Science and Technology in Ghana. It offers consultancy services at all levels of manufacturing industry: members of the Centre seek clients by frequent visits to factories and workshops.

Another variant is the Hajj Research Centre at King Abdul Aziz University, Jeddah, Saudi Arabia. The Centre provides an excellent example of target-oriented basic research and due to an acute lack of research on the ecology of Mecca and the problems of Hajj, the local consultancy firms line up to use the results of the Centre.

An alternative approach is to exploit the new regional universities in the Muslim countries for technological development. As provincial universities are generally newer and more recent than metropolitan universities, they should be more flexible and responsive to innovation. Programmes among staff and students could be developed to stimulate research related to the local economic and social needs of the region, through courses, workshop, degree projects and postgraduate work.

Under what conditions can we expect such efforts to be successful in creating genuine linkage between the universities and industrial organisations? In most cases, these arrangements can be seen as a modified form of a research utilization system in which one organisation attempts to develop its own links with clients rather than participating in a nationwide system in which similar efforts of many organisations are coordinated by a government agency. A number of factors are likely to affect these relationships. A key factor is likely to be the extent to which the university accurately defines and locates its potential clients. The researcher himself will probably be the last person to engage in this kind of activity due to lack of time and possibly lack of appropriate communication skills. This suggests that universities may need to hire and train their own 'extension' specialists who would develop skills of communicating to businessmen information about the types of services the university could provide for them. In addition, the specialist would have to learn how to convey the needs of industrial firms to the university researchers. A two-way transmission of information will be essential to the success of such a project. This type of approach can be expected to be most useful in dealing with numerous small firms whose research needs are fairly similar.[26]

An alternative to university research institutes are government-backed laboratories and research organizations. On the whole, the linkage with industry in some of these organizations is better than the linkage between industry and universities. Indonesia's 28 government-backed laboratories, for example, are engaged in industrialization research, environment and resources development research, and basic science activities.[27] Pakistan has recently begun to lay emphasis on scientific and technological research by allocating increased finances for the promotion and establishment of scientific research organizations. In 1968, there were 180 research establishments backed by the government, as well as universities and autonomous funding institutions actively engaged in linkage activities.[28]

As most of the government-backed research organizations are engaged in industry-orientated research, these institutions are in a good position to bridge the gap between research, and its consumption. However, the success stories are few. There are many research centres in the Muslim world with sizable completed projects awaiting exploitation by entrepreneurs.

As research problems are common for many Muslim countries, the suggestion of Majid Rahnema that the Muslim countries should 'create a pool of their own brains in order to utilize in common their highly qualified people instead of losing them to developed countries' deserves serious consideration.[29] The idea is to establish a kind of central pool of all the qualified personnel belonging to Muslim countries from which any member country could draw. Coupled to this idea is the suggestion to set up joint projects for R & D which would pool available resources to solve urgent research problems. On this line the Regional Cooperation for Development (RCD) between Pakistan, Iran and Turkey has had reasonable success.

Of particular consideration for joint R & D are the scorching problems of drought in the Sudano-Sahellian zone and the associated need for research in the field of hydrology and other areas. Studies of desertification affecting the arid zone regions and solar energy deserve equal attention. Basic research work on solar energy is proceeding at the University of Dakar, Senegal, while similar work is being duplicated elsewhere in the Muslim world.

The effect of protein deficiency on mental and behavioural development, malnutrition and retarded physical development of the young may be other fruitful areas of joint effort; so also may research on the locust and the study of the eradication of the tsetse fly which would open up vast areas of Muslim Africa for settlement and agricultural development. Joint research and cooperation would eliminate duplication of effort on common problems and may ensure the maximum use of existing facilities.

For the purpose of joint research between various Muslim countries we would suggest the establishment of research centres, on somewhat similar lines as those suggested by Carl Djerassi,[30] of an internationally recognized standard of excellence, in various Muslim countries. Such research centres would tackle target-orientated basic research problems common to the Muslim world; and be manned and directed by an international cadre of Muslim research fellows. The advantages of such centres would be many; Djerassi lists a few:

1. By demonstrating in a few developing countries that internationally recognized basic research centres can be created without waiting for the logical education and technological development of the country, an example is created, which (with possible modifications) may prove useful in other developing areas of the world.

2. By selection a research field which offers ample opportunity for fundamental research and yet may have eventual practical technological consequences, the possibility of generating new industrial development is presented.

3. By selecting a research field with a maximum eventual multiplication factor (for example, chemical research, which requires microbiological, entomological, and pharmacological evaluations, and leads eventually to clinical and veterinary applications), a fairly rapid and yet logical broadening in the scope of the research effort can be effected.

4. In countries where scientific research offers no status symbol whatsoever, a beginning will be made in this direction so that eventually scientific research may become a desirable career in that society.

5. An opportunity will be provided in the particular country for advanced training in certain research fields without the necessity initially of sending promising young students abroad. Once they have been exposed to serious research in their own country, subsequent foreign training (even in another field) becomes more

meaningful and a return to the home country much more likely, because they will realize that research of an internationally accepted standard can be performed in their home country.

6. One factor in the complicated brain-drain problem — the non-availability of research facilities — will be partially eliminated in certain fields.

7. To raise the overall scientific level of a country or even of one university is difficult and time-consuming. To create selected centres of excellence in a few fields is easier and their existence, in turn, frequently creates a stimulus in many other areas. Furthermore, the local 'image factor', though of limited value, should not be ignored, since it is easier to point to a concrete and operating entity (for example, the Rice Institute in the Philippines) than to a statistically significant reduction in the literacy rate of a country.[31]

Of course, the problems associated with setting up such research centres would be many, and several would be particular to the countries selected. However, the problems should be seen as challenges to be tackled and overcome.

Development of 'pools of Muslim brain' and international target-orientated pure research centres does not, needless to say, reduce the importance of national universities as centres for research. We believe that no hope for development is real without the universities in the Muslim world becoming dynamic, thriving, centres for ideas as well as research. To achieve this the existing university structures may have to be modified or renovated or new research-orientated universities established.

NOTES

1. C. Freeman, 'Measurement of Output of Research and Experimental Development — A Review Paper' (UNESCO Statistical Reports and Studies, No. 16, Paris, 1969).
2. M. J. Moravcsik gives a detailed analysis of research literature. See his *Science Development. The Building of Science in Less-Developed Countries* (PASITAM, Bloomington, 1975), pp. 119-36.
3. H. Roderick, 'The Future Natural Sciences Programme of UNESCO', *Nature* 195, 215 (1962).
4. Harvey Brooks, letter, *Minerva* 10, 327-8 (April 1972).
5. C. Freeman and A. Young, *The Research and Development Effort of Western Europe, North America and Soviet Union* (OECD, Paris, 1965), p. 66.
6. 'Third World: Science and Technology Contribute Freely to Development', *Science* 189, 770-776, 4205 (1965).
7. Mahbub Ul Haq, Wasted Investment in Scientific Research', in W. Morehouse (ed.), *Science and the Human Condition in India and Pakistan* (Rockefeller

University Press, New York, 1968). Haq points out that in the mid-sixties, ten times as much money was spent in Pakistan on nuclear research as on research on natural resources.

8. D. de Solla Price, 'Nations can Publish or Perish', *International Science and Technology*, October 1967; and A. B. Zahlan, 'The Science and Technology Gap in the Arab Israeli Conflict', *Journal of Palestinian Studies* 1, 17-36 (Spring 1972).

9. M. J. Moravcsik, 'Science and Technology in Natural Development Plans: Some Case Studies'. Paper submitted at the Symposium of 'The Incorporation of Scientific and Technological Considerations into Development Planning', Agency for International Development, Washington, 30 April 1973.

10. M. Ul Haq, 'Wasted Investment in Scientific Research.'

11. N. Stepan, *Beginnings of Brazilian Science. Oswaldo Crux , Medical Research and Policy*, Science History Publications (Chapter on Science in a Developing Country. 'Science Policy Issues'. (To be published).

12. Moravcsik, *Science Development*, pp. 115-6.

13. Moravcsik, 'Science and Technology in Natural Development Plans.'

14. Ibid.

15. See Peter Newmark's assessment of 'Overnight Transplants' in 'Iranian Transplant', *Nature* 261, 358-9 (1976).

16. Quoted in *Science* 189, 4205 770-776 (1975). Although the statement refers to the developing countries in general, it is particularly relevant to the Muslim countries.

17. Pace, numerous papers by D. de Solla Price and others.

18. Diana Crane, *Invisible Colleges* (University of Chicago Press, 1972).

19. Abdus Salam, 'The Isolation of the Scientist in Developing Countries', *Minerva*, 4, 461-65 (1966).

20. Moravcsik, *Science Development*, p. 83.

21. For a discussion of the importance of science journals in the creation of self-sustaining scientific infrastructures, see George Bosalla, 'The Spread of Western Science', *Science* 156, 611 (1969).

22. For a general survey of libraries in the Muslim Countries, see Anis Khurshid *et al.*, 'Fact Sheets on Libraries in Islamic Countries' (University of Karachi, 1974).

23. S. J. Haider, 'Science Technology Libraries in Pakistan', *Special Libraries* 65, (10/11), 474-8 (1974).

24. E. Turkcon, 'The Limits of Science Policies in a Developing Country: The Turkish Case, a Study Based on the Experience of the Scientific and Technical Research Council of Turkey', *Research Policy* 2, 336-63 (1974).

25. Philip Ward, 'Indonesian Libraries Today', *UNESCO Bull. Lib.* 29 (4), 182-7 (1975); and A. G. Myatt, 'Scientific and Technical Information in Indonesia — Problems and Prospects', *Bll Review* 1 (2), 52-6 (1973).

26. Diana Crane, 'An Inter-Organisational Approach to the Development of Indigenous Technical Capabilities. Some Reflections on the Literature' (OECD (CD/TI (74) 31), Paris, 1974), p. 15.

27. Leon Press, 'Scientific Research Institutions in Asia', *Impact of Science on Society* 19 (1), 25-51 (1969).

28. M. A. Khan, 'Education Research in Pakistan', *Impact of Science on Society* 19 (1) 85-93 (1969).

29. Majid Rahnema, 'Iran: Science policy for Development', *Impact of Science on Society*, 19 (1) 53-61 (1969).

30. Carl Djerassi, 'A High Priority? Research Centres in Developing Nations', *Bulletin of Atomic Scientists* (January 1968), pp. 22-27.

31. Ibid., p. 24.

10 PATHS OF ACADEMIA

Education has a vital role in the supreme effort to preserve a society's cultural identity and historical legacy. It supplies its members with a coherent frame of reference for the values held highly by that society. Here the concerns of the Muslims are not unique. Other societies and other cultures also have, from time to time, resisted the subversive efforts of education by alien civilizations on their cultures:

At the Treaty of Lancaster, in Pennsylvania, anno 1744, between the Government of Virginia and the Six Nations . . . the Commissioners from Virginia acquainted the Indians by a Speech, that there was at Williamsburg a College with Fund for Educating Indian youth; and that if the Six Nations would send down half a dozen of their young lads at that college, the Government would take care that they be well provided for, and instructed in all the learning of the White People.

[The Indians' spokesman replied:]

'We know that you highly esteem the kind of Learning taught in those Colleges, and the Maintenance of our Young Men, while with you, would be very expensive to you. We are convinced, therefore, that you mean to do us Good by your Proposal; and we thank you heartily.

'But you, who are wise, must know that different Nations have different Conceptions of things; and you will therefore not take it amiss, if our Ideas of this kind of Education happen not to be the same with yours. We have had some Experience of it; Several of our young People were formerly brought at Colleges of the Northern Provinces; they were instructed in all your Sciences, but, when they come back to us, they were bad Runners, ignorant of every means of living in the Woods, unable to bear either Cold or Hunger, knew neither how to build a Cabin, take a Deer, or kill an Enemy, spoke our Language imperfectly, were therefore neither fit for Hunters, Warriors, nor Counsellors; they were totally good for nothing.

'We are however not the less obliged by your kind offer, tho' we decline accepting it; and, to show our grateful sense of it, if the gentlemen of Virginia will send us a Dozen of their Sons, we will take care of their Education, instruct them in all we know, and make Men of them.[1]

In most Muslim countries, modern educational establishments have developed either in the aftermath of colonialism or from conflict with traditional systems of education. The modern educational systems, in a great many cases, have been transferred *ex occidente* in a way rather analogous to the transfer of technology, with the basic assumptions of the occidental education system as well as the traditional occidental dogmatic conceptions. By and large, the educational philosophies of this system are corrosive to the value systems of the Muslim countries. For one thing it divides education into two watertight compartments: the 'secular' and the 'religious'. In Islam this division is unknown. This is why after decades in operational form, the modern education system is producing clear signs of unresolved ambiguities and a manifest alienation from the cultural roots of the society.

In some Muslim countries, the occidental education system was forced on the indigenous population by the occidentalized ruling group. Turkey suffered much at the hands of the Young Turks and Ataturk especially; Iran from the two Pahlavi monarchs.

In other Muslim countries, such as Pakistan, Algeria, and Malaysia, occidental education has been a colonial legacy. With occidental aggression and colonialization, coupled with the Christian missionary and imperial trading companies, came the occidental education system as a powerful instrument to social change. The goal of this import was simple and direct: to 'civilize the native', to change his way of life. It was a package deal containing the deep-rooted seeds of Islam-Christendom conflict and the moral arrogance and 'rational materialism' of nineteenth-century Europe. On the part of the receiving culture there were two contradictory reactions. There were those traditional scholars who declared that even to learn occidental languages amounted to a kind of surrender. And there were those who could see the salvation of the colonized Muslims only in terms of total acceptance of occidental education.

These two viewpoints are still predominant in the Muslim world. In most Muslim countries the two philosophies exist side by side as divergent educational systems. However, the occidentalized educational institutions are dominant, enjoy support of the government and high status and provide the output for the better jobs in the country.

The occidentally educated Muslim can seldom be anything but a mutant. The reason for this is quite simple. The Muslim's dialectic of being—of his philosophy, of his sense of reality—can never be reduced to a scheme, a skeleton to be filled in by the material of some alien spiritual and value system. It has its roots deep in a religious outlook that has moulded the trials and passions of a singularly religious people with a keen sense of social consciousness.

The intellectual framework of a Muslim is complete and consistent within itself. Here there is room for the exercise of reason, for the awakening and disciplining of emotions, for the exertion of sense-perception, and for reflection on deep, inner feelings. Within this framework is the systole and diastole of the universal heart. Here experience is a unitary process in which human beings participate with all capacities—feeling, thinking, willing, interacting—and all value systems—factual, aesthetic, moral, spiritual. Without his metaphysical world, the Muslim is a body without a heart. It would be disastrous to remain blind to this fact.

In contrast, the occidental education system is in no need of metaphysics. This system depicts life in all its material richness and variety, but leaves it without a setting, and as a result without a meaning.

The Occidentalized Institutions

And what of the product of the occidental education system? In comparison to the true occidental student he is a very bad specimen. The most notable feature is a (often complete) lack of critical, analytical, or down-right argumentative attitude that is part of the process of learning. He is also far removed from the culture of his parents and often develops a defensive superiority complex.[2]

Those who have taught this variety of student—in the Occident as well as in the Muslim countries—are well aware of the 'guru attitude'. This attitude reveals itself in the dictum: the teacher is always right, even when he is known to be blatantly wrong. This attitude plays a great part in subverting critical and analytical faculties as well as the use of imagination. The pupil of such a system not only ends up with a poor, out-of-date collection of scientific knowledge but he is also quite incapable of thinking for himself.[3]

This system of mediocrity is relentlessly perpetuated. At some stage in time, the guru worshipper himself becomes the master guru. How could his students possibly develop the reflective, investigative, and experimentally orientated attitudes that go to make the 'academic spirit'? How can the product of such a system generate new ideas, question occidental ones, and instil the kind of confidence that is needed to solve the most pressing and intractable problems of the Muslim countries?[4]

The occidentalized institutions of learning in the Muslim world have failed to import innovative and imaginative methods and techniques of education relevant to and realistically geared to local needs. Many of these institutions are so alien that there is hardly any interplay between them and the societies for which and with which they were created. Their interests are always marginal and their philosophies completely alien.

Occidental universities tend to exemplify middle-class European culture, and the norms and values that go with it. Yet to the students in Muslim countries the experience of the European middle-class is completely absent, quite unknown. But they have strong motivation to succeed: this motivation often tends to be related to socio-economic and survival needs—including employment, social mobility and power. The cultural gap between the home background and the universities generates a great deal of insecurity and tends to emphasize the material aspect of education at all levels. The end products are graduates of 'distinct personality types of dissimilar and often conflicting cultural orientations'.[5]

The high level of illiteracy has meant that the society could not provide any feed-back to these universities. This has resulted in a lack of understanding, on the part of the university, of the make-up of the society as well as its needs. Graduates from the occidentalized universities have little regard for the traditional culture and even less sensitivity to their environment.

Teaching is limited to the requirements of examinations, and the students' primary aim is to pass examinations—by hook, or in some cases, by crook.[6] An example from Pakistan:

> ...The adaptation of unfair means [is] common and has now increased to such an alarming extent that no result is a sure indication of the candidate's performance or ability. The disease affected many students, their parents, a fair number of teachers and quite a few examiners. As the sole aim of education became to pass examinations, there was continuous pressure that they should be made as easy as possible. Any effort to raise the standards was resisted. After every examination the newspapers are full of complaints by angry students criticising the stiffness of question papers. Examinees even object to 'unexpected' questions. If they do not like a question paper, they walk out of the examination hall and disrupt the entire

examination by moving from centre to centre and forcing those who are not dissatisfied with the question paper and want to continue with the examination to leave the examination hall. If too many candidates fail, the examining body is the target of harsh criticism.[7]

A particularly acute problem faced by the universities concerns the training of scientists, technologists and teachers. On the professional level there are two related difficulties: a shortage of qualified teachers of science and technology and inadequate teaching material in the indigenous language. Those who can teach can do so only in a foreign language, for they themselves were either trained abroad, or if locally, through a foreign medium. In countries formerly under British control, English is the language of science and technology, but in countries where the colonial rulers were French or Dutch, these are the languages of instruction. Textbooks, therefore, have to be translated to be within the reach of the students. Teachers, trained abroad, need time and sustained effort to be able to teach through the medium of the natural language of the students.[8]

On a social level, prejudices exist against technical work and training of all kinds and in favour of purely academic work. This is largely a legacy of colonial days when the main aim of local education was to provide clerks for the colonial administration. However, with the present trends in the import of technology, the prestige as well as wages of technicians and mechanics have gone up. Much more emphasis is required on the training of technologists; in Arab countries, for example, only 15 per cent of the student population is in technical institutions.[9]

A great deal of reliance is also placed on sending students abroad for training. This is not merely at postgraduate level but also quite commonly at the undergraduate level. After sampling life in the Occident many come to identify with it, and often fail to return, or only return for the time necessary to fulfil their contracts. The aim in sending these students to the Occident is to provide a crop of technologists as well as managers, executives and teachers. However, the net result is often the loss of talented students to the Occident. Those students who do return can seldom teach their subjects in local languages. Often the same level of training could have been provided in a nearby Muslim country where the student would at least have been in a familiar cultural milieu.[10]

Those with occidental qualifications usually fare better in employment than home-grown technicians.[11] They have better jobs and a higher status. But these qualifications are often ill suited for the needs of the country and provide it with little benefit. The relevance of many Ph.D. projects is coming under attack in the Occident for irrelevant and narrow specializations. One does not have to stretch one's imagination to realize their unsuitability to the developing countries.

In the end, the benefit of sending students abroad for training, for a Muslim country are more apparent than real. We would suggest that before sending students abroad for training the Governments of the Muslim countries ask themselves:

1. Is the student going to get the right training, at the right level, for the job he will be doing on his return?

2. Is it necessary for him to go to the Occident? Can he not obtain the required training in a nearby Muslim country?

3. Will there be a job for him when he returns?

4. Is it necessary for him to conduct a three year research project for a Ph.D.? Would a Master's course be sufficient?

The Traditional System of Education

The traditional system of education, in comparison to the occidentalized version, works well within its limited frame of reference. It is relatively deficient in quantity and quality of books, teaching and general conditions of study, classroom facilities and teachers. The teaching methods are usually obsolete, teaching styles are antiquated, and overall progress of the students is slow. Syllabuses do not normally include subjects like science or modern occidental languages. Yet the system manages to achieve what it aims for: to give a sound religious and cultural orientation and provide a classical (Arabic) education as well as an education in classics, to produce individuals who are at one with themselves and the society they inhabit.

The product of this system is largely responsible for preserving the cultural heritage of Muslims. The very same people laid the foundation of the 'Freedom Movements' in the Indian subcontinent, defended the Khilafah in Turkey, liberated Algeria and Tunisia from the French, forced the Dutch to withdraw from Indonesia, uplifted the Arabs after their defeat and disgrace at the hands of Israel. Few would doubt the dedication of the traditionally educated Muslims for the service of the *Ummah*.

Contrary to popular belief, the traditional education methods do not rely entirely on 'obedience and memory'. This is only needed when the children are learning to memorize the Qur'an which has to be preserved not just as a written text, but also in the hearts and minds of Muslims. Memorizing the Qur'an is an established tradition, based on learning and experience, with its own methods and techniques which naturally involve some reliance on memory. However, memory plays an insignificant role when other aspects of religion are taught. As we pointed out in Chapter 2, within its framework Islam includes a sophisticated awareness of reason in strengthening religious traditions and a critical scholarly attitude. These play an important part in teaching the wisdom of the Qur'an, the *Seerah* (life) of the Prophet, *Hadith* criticism, techniques of *qiyas* ('analogical reasoning'—the fourth source of Islamic law), Islamic legislation, essentials of Muslim jurisprudence, the role of the individual in society, and so on. Thus the very process of studying Islam instils the awareness of reason as well as criticism and scholarship. This is why the traditional education has made a more original contribution to Muslim societies and has produced many internationally celebrated scholars even in the post-colonial era. In contrast, the contributions of the occidentalized Muslim scientists and scholars have been insignificant both on the national and the international scenes. (After all how many Muslim scientists are there of international repute?) The real merit, however, is in the system; and not in the individuals.

This is why we believe that the aims of the principle of domesticity can really be served best by the products of the traditional schools. Unlike the occidentalized elite, they really have the interests of the local community at heart.

All this is not to say that the traditional system of education is perfect. For one thing, it is not completely in touch with today's realities (but, then, neither is the occidentalized system!) For another, it is not too keen to adopt what might be considered improvements in teaching methods. And yet another, it is not equipped to impart scientific and technological education. But despite all this the traditional education system is ideally suited to represent the principle of domesticity which we consider to be vital for any strategy of self-reliance. However, it is in need of some serious modification.

The most important role the traditional sector can play, in our opinion, concerns the creation of a 'scientific awareness' amongst the Muslim populace. It is not, as is commonly believed, against science and technology as such, although some traditional scholars may well be against the vulgar variety of occidental science and technology.

Most proposals to educate the public about science and technology are based on one assumption: that the Muslim societies must learn English. However, it is neither easy nor possible in some cases for

Muslims to learn English. Arabic being the language of the Qur'an, non-Arabic speaking Muslims give Arabic a top priority for their second language. Furthermore, in the minds of many Muslims, English is the first step towards occidentalization. As such the use of English as a medium for creating awareness of nature is very much a non-starter.

The traditional scholars, however, are closest to the grass-roots of Muslim societies. As such they are in a unique position to teach basic science in the natural language. But more than that, the traditional system can relate science to the background of the people. The traditional backgrounds of Muslim societies are fully aware of the laws of nature. A 'scientific awareness', in the sense of awareness of nature, can thus easily be instilled in Muslim societies, if these laws are explained within their surroundings and in their natural language. The role of astronomy in their patterns of prayer and fasting; the use of algebra in the Muslim laws of inheritance, and amongst others, surface tensions, charcoal heating, wave behaviour, musical instruments, beats and resonance, body biology—are but a few areas which can be related to the background of Muslim societies. With a heritage of science as rich as that of Islam, relating science to tradition is not a difficult task. The Muslim populace could become much more aware of natural laws if the everyday behaviour of these laws is explained in the people's natural language. The mathematical description would then naturally follow.

Perhaps the most important reason for using traditional scholars to teach basic science is that the view of reality on the part of the teacher and the pupils and students—who would mainly be from the rural areas—is one and the same. Replace the traditional teachers with the occidentalized variety and you will find that the teacher and pupils have different understandings of reality. The resulting communications gaps are formidable.

We like to think that science properly taught will give us understanding of the 'real' world. Which real world? People everywhere believe they know what is real and what is not, and most of them fit their knowledge of reality into a coherent structure of ideas and values that serve their needs. Indeed each 'reality' left to itself may serve its own inventors very well, but problems may arise when a culture changes too rapidly or when there is a transition and superposition of cultures. A number of these problems can be brought into perspective by examining the teachings of science in one or another of the non-Western developing countries. In these countries science teaching, for the most part, has little or no connection with the intellectual life of the community outside the school and, therefore, what is taught receives little reinforcement from that community.[12]

The intellectual leadership in rural areas of many Muslim countries is provided by the traditionally educated Muslims. To use them for teaching science is an obvious way of bridging the gap between the school and the community.

Some of the distinct advantages of using the traditional educational system for primary science teaching, and creating an overall scientific awareness amongst the Muslim populace are:

1. The present purpose of the traditional education systems is to provide what is loosely called 'religious education'. The children are sent to the *Madrasa* or *Pondoks* at a very tender age, the age when a spirit of enquiry and experimentation can be created in them. Teaching science, as well as the Qur'an, tradition, and *Usul al-Din*[13] in the *Madrasa* will enlarge the scope of traditional education as well as place science in a concrete setting. Here science will be taught not only in the natural language but will also be related to the background of the children. All the phenomena that their parents may have explained as miracles, can now be explained scientifically in the natural language.

2. As the traditional learning sector is looked up to in rural societies, and as it comes—more than anybody else—directly in contact with the Muslim populace, it can initiate table discussions on scientific events and topics and thus pave the way for a sound scientific awareness.

3. It is commonly accepted that the dissemination of knowledge amongst the rural population is only possible if it is done in a natural common language. The traditional system, using natural language, will ensure a real and rapid dissemination of knowledge.

4. Properly managed it can be a great force in bringing unity to a community.

5. Discovery and innovation will become much more of a reality —that is, discovery and innovation that are necessary and relevant to the society.

All this, of course, can only be done if traditional educational systems are given the importance that they deserve and, where they have gone into hibernation, revitalized. The condition, in our opinion, of Muslim countries could be better improved by updating and upgrading the syllabuses of the traditional education systems, and giving them the financial resources and training that is needed to reactivate them.

The New Universities

There are marked differences in quality among the universities of the Muslim world, ranging from good to ordinary to downright mediocre. There are, however, no real first-rate universities anywhere in the Muslim world. Although there is a gradation in quality, there is no precise way of classifying them. We have the traditional Islamic universities that stand apart from the occidentalized imports which were established under the colonial administration or immediately after independence. Then there are more recent regional universities which have tended to concentrate on technical education.

The old-established Islamic universities, like the Al-Azhar in Cairo and Qarawiyyi in Fez, teach traditional Islamic subjects: the graduates of these institutes tend to be well-equipped, despite the odds, in traditional knowledge of the culture.

The now well-established occidentalized universities teach a whole range of subjects from pure and applied sciences to social sciences, arts and humanities. Some of these universities are: University of Punjab in Pakistan, University of Rabat, University of Tehran and University of Malaya.

The newer universities are regional in character and have evolved only over the past ten to fifteen years. Most of these universities are technically biased; but at the same time they are also, to some extent, Islamically orientated. Examples of the new universities include the University of Islamabad, Pakistan; Ahmed Bello University, Zaria, Nigeria; King Abdul Aziz University, Jeddah, Saudi Arabia; University of Kuwait; the National University of Malaysia; and the *Centre Universitaire* at Oran and Constantine in Algeria. These universities have faculties of engineering and science, technology and medicine, and some even have faculties of arts and letters, but most have strong faculties of Islamic studies.

What makes the new universities different from the old occidentalized universities is not just their vitality, but also their belief in and efforts to revive Muslim scientific traditions. The teaching and research standards at these universities are good and improving rapidly. However, a straightforward comparison with European or American institutions is difficult as well as meaningless.

We believe that the real contribution of the new universities is the re-thinking they have forced on themselves. Along with an increasing awareness of local problems, and a keen desire to solve them, there is also developing a questioning attitude towards governments and other institutions and an effort to re-think their policies and approaches. The first thing to be affected by this attitude has been the relationship between higher educational institutes and the governments. As a result, higher education is being diversified, and new institutions—quite different from those of the Occident—that reflect the local culture are being created. This greater flexibility has led to interdisciplinary approaches to research problems, there being a greater tendency towards the task-force concept.

The research being carried out at the new universities is, by and large, geared to the needs of the country. It relates to teaching, and sometimes, to the problems of local industry.

The links between industry and the university are being forged—often by the staff *going out* to industry to offer assistance. As such, there is more reliance on local resources and expertise and real effort to develop indigenous scientific and technological capacities. Coupled with the move towards industry, there seems to be a greater co-ordination between applied and pure research.[14]

A unique feature of the new universities is the future-orientated thinking that they have developed. After examining a number of assumptions about tomorrow's society, the educators of these universities have been speculating imaginatively about the structures of schools, universities and institutes that will best suit their societies tomorrow. Many experiments in curriculum design have been carried out and emphasis is on curricula which create a greater awareness of local problems. Many modifications of the uncritically accepted curricula of the occidental universities, adopted by the local universities, have led to more indigenously oriented courses.

All these are signs of dynamism, and lively, critical activity. However, the ideas initiated and generated by the new universities have not permeated to other institutions of higher learning. Where older establishments have come in contact with new ideas, the reactions have often been hostile. So, in general, universities in the Muslim world continue to see their roles as nothing more than teaching machines. Scientific and technical knowledge is obtained from the Occident and passed on to the students in a raw, unprocessed form. Students are spoon-fed and encouraged to memorize and cram facts on the assumption that a university's *raison d'être* is to prepare students for examinations and paper qualifications. One of the best features of Islamic education was the personal contact between teacher and student in the form of the tutorial system. Today, the only form of communication between students and teachers is the impersonal lecture. Seminars and academic discussion groups are all too rare. In short, the teachers have very little opportunity for the discovery and development of talents and potential of their students.

In such circumstances, we feel that there is an urgent need for policies that broaden the outlook of the universities and bring back the missing academic spirit to these fossilized institutions.

Broadening University Horizons

The university has a role somewhat similar to the higher nerve centres of the body in abstracting and combining, analyzing and synthesizing, assessing and reassessing all the other information of the society — in short, it is a thriving *thinking* system. This dimension is conspicuously missing in the universities in the Muslim world.

As teaching institutions the universities like the schools rely a great deal on paper examination to test the attainments of their students. These tend to cram too much, and that leaves little room for independent thought and judgement. It has been said, and there is wisdom in the saying, that the best university education is that which teaches the least. This may be interpreted to mean traffic of ideas between teacher and pupil is more important than formal lecturing.

But the pressure of numbers and the force of old habits in administrators and professors tend to place students almost beyond the personal influence of their seniors. The tutorial system and the small seminar are still novelties. On the other hand lecture halls and laboratories are getting fuller, so full that at times amplifiers are used for the benefit of 'listeners' in the corridors. How many lecturers, engrossed in this tedious routine, realize that knowledge is not 'static', but constantly changing, even in the natural sciences? How many have the time, inclination or mental agility to revise their lectures at least every year? How many social scientists and men of letters do profit from and adopt the scientific method? How many realize the loss to themselves and their speciality which results from the lack of close contact and exchange of views with the young?[15]

Universities, to the Muslim way of thinking, have functions much broader than just that of teaching. In the total intellectual enterprise of human society, universities are obviously the primary repositories of available knowledge, the centres for reflection on and analysis of that knowledge, and the generators of new ideas. These were the functions performed by the great universities of Baghdad and Cordoba, Granada and Samerkand. And these are the functions the universities of the Muslim world must perform today.

As a repository of available knowledge, universities must have facilities for traditional scholarship and access to knowledge and library storage of everything that humanity has known or said or thought or done. Within the framework of traditional scholarship there is an area for processing and analysis. Only after analysis can this organized knowledge be filtered and communicated to the next generation. Unfortunately, in the Muslim world, scholarship amounts to little more than collection of facts, parliamentary records, cabinet proceedings, minutes of extraordinary meetings and the like. Scholarship of science is unheard of.

Along with traditional scholarship there is also the acute need to carry out basic studies in natural, biological and social sciences. This knowledge has to be transmitted for the use of the society in writing by publishing papers and books, in consultancy work for firms and industry, and to the community, business and government through appropriate channels. As such, alongside this traditional disciplinary education, the universities must establish social and applied R & D centres. This is a job only the universities can do. What other institutions could possibly have the required depth of knowledge, breadth of vision, the information and library facilities, the laboratories for testing and analysis, and the time to look at the needs and concerns of the *whole* society rather than some narrow sector of profit or bureaucratic concern?[16]

One responsibility of the university towards the community is in the field of adult education. The adult illiteracy rates in the Muslim world verge on the ridiculous figure of 90 per cent, yet the universities have played virtually no part in adult education. We believe that in this matter the university has not only a responsibility, but also an opportunity. We have already mentioned the role the traditional scholars can play in creating a scientific awareness among the rural population. The universities have to cooperate with and equip the traditional scholars for this role. Also there is the role of radio and television in preparing adult education courses to be considered. The need here is for many small schemes continually operating rather than grandiose schemes with unattainable ambitions working on a 'once and for all' basis. The aim must be, as A. L. Tibawi has pointed out, to make all mothers progressively literate and to give them at the same time essential training in child rearing and home management. In the second place, semi-skilled labourers in the fields and factories must be made effectively literate. Finally, those members of the armed forces who have missed school must be made targets of literacy schemes. To help adults in today's Muslim societies attain and retain literacy is an urgent as well as a difficult task. Its success, however, must be linked with the social, economic as well as national needs to employ usefully the adults liberated from illiteracy.

In their traditional role as teaching institutions, the universities in the Muslim world have a great deal to learn and unlearn. In particular, we suggest that:

1. Less emphasis should be given to memory and exam-orientated learning. Students should be taught to develop their intuition, insight, and creativity as well as their faculty for rational thinking. They should be encouraged to work creatively without fear of authority.

2. Teachers and professors should learn to recognize the extent to which they have been 'programmed' by the occidentalized system of education. They should re-examine their value-orientation and especially the inherent values and cultural bias in what they teach.

3. Both professors and students must guard against the false application of the 'scientific method'.

4. University administrators and decision-makers must accept that the best way to make decisions is not through the 'mechanistic method' but via consultation, dialogue and debate. All this is facilitated by trust and respect; while excess deference to 'expertise' and heavy reliance on foreign 'consultants' impede good decision-making. 'Within the atmosphere of trust and respect, controlled conflict and anxiety can be expected to divulge important insights for decision-making and for life itself.'

5. Professors and students have a right to hold divergent views of academic issues without being subjected to explicit or implicit sanctions.

Finally, we come to the responsibility of the universities in shaping the future of the society they serve. It is the responsibility of the universities to grasp the future, to provide information and policy guidelines in the seminal creation of world-changing inventions or paradigms or critiques—such as cybernetics and socio-biology, atomic energy and feed-back theory, ecology and the concept of 'limits to growth'. These are not rising 'disciplines' but 'movements' which can have profound effects on Muslim society and culture. In this respect, universities have to be not just seats of learning but also citadels of wisdom and centres for all kinds of policy guidelines.

The universities control the reins of the future of the Muslim societies; their responsibility of guiding Muslim societies in the right direction, away from occidentalism and towards Islam, cannot be underestimated.

NOTES

1. Benjamin Franklin, 'Remarks Concerning the Savages of North America', London 1784, quoted by Thomas Lambo, 'Relevance of Western Education to Developing Countries', *Teilhard Review* II (1), 2-4 (1976).
2. See Ziauddin Sardar and Dawud Rosser-Owen, 'Science Policy and Developing Countries', in I. Spiegal-Rosing and O. de Solla Price, *'Science, Technology and Society'* (Sage Publications); and Moravcsik, *Science Development.*
3. A Pakistani graduate of science, for example, compares poorly with a British student with two science 'A' levels.
4. Sardar and Rosser-Owen, 'Science Policy and Developing Countries'.
5. *Islamic Education in Malaysia* (Ministry of Education, Malaysia, 1976), p.2.
6. S. H. Alatas, *Sociology of Corruption* (Donald Moore, Singapore, 1968).
7. Ishtiaq H. Quarashi, *Education in Pakistan* (Ma'aref, Karachi, 1975), pp. 55-6.
8. A. L. Tibawi, *Islamic Education* (Luzac, London, 1972), pp. 214-22.
9. Ibid.
10. Sardar and Rosser-Owen, 'Science Policies and Developing Countries'.
11. A. B. Zahlan, 'Science in the Arab Middle East' (1967, unpublished); C. Nadar and A. B. Zahlan, *Science and Technology in Developing Countries* (Cambridge University Press, 1969); R. Scalapino, *Asia and the Road Ahead* (University of California Press, 1976); and others.
12. P. E. Dart, 'Science and the Worldview', *Physics Today*, June 1972, pp. 48-54.
13. 'The Rules of Religion.'
14. Alexander Dorozynski, 'Eden with Oil Wells', *Nature* 257-80 (1975); and Tibawi, *Islamic Education.*
15. Tibawi, *Islamic Education*, p. 219.
16. In this respect, see the excellent paper by John Platt, 'Universities as Nerve Centres of Society', in *The Future as an Academic Discipline*, Ciba Foundation Symposium, 36, North Holland, Amsterdam, 1975.

11 THE FUTURE

And so to the future itself. The future which is the foundation of the hopes and aspirations of the Muslim *Ummah*, as well as its fears and anxieties. The future that depends not just on the past of Muslim societies but also on the policies they pursue today, as well as the vision they nourish of the future.

In all honesty, the reactions of the Muslims have been nothing less than traumatic to any suggestion of long-run planning and futuristic thinking. There are strong tendencies to withdraw to positions of nostalgic irrelevance, to cultivate dreams of bygone ages, and to refuse to think of the future and be part of new solutions. For some unknown reason the Muslims have taken the sublimely bad advice of Mr Gladstone: 'You cannot fight against the future.' But *you* can!

To try and predict the future of the Muslim *Ummah* thirty, forty, fifty years hence would be a hazardous undertaking. We have no sympathy with astrology. The Muslim sage Sadi Shirazi once said, 'the wise try to learn if even wise words are written on the wall'. To a large extent policy-making involves reading the writing on the wall. And wisdom is a vital ingredient for planning as well as for implementing policy. Whatever the doubts, it is necessary to have some understanding of what lies beyond, what would be the possible outcomes of certain policies, whether the understanding is gained by projection of possible future developments or by a mixture of knowledge and informal speculation. Some understanding of what could happen a few decades from now if the present policies and trends continue, would help our discussions of the right policy mix for the future.

Let us illustrate the point. Consider the case of Kuwait whose economy is based entirely on the sale of oil. (However, the discussion that follows can also be applied, with due modification, to other oil-producing Muslim countries.) Almost up to the Second World War, Kuwait was a poverty-stricken sheikdom, without reasonable supplies of water, food, shelter and other necessities of life. Now the oil revenues have changed the situation. However, oil as a source of income cannot last forever; once it has been pumped it has gone forever. The current estimates of the present oil reserves of Kuwait (65 billion barrels) along with the data on current pumping rate (2.6 million barrels per day) suggest that the oil wells will dry up in fifty years.[1] What happens when the oil runs out? Do they return to square one? At present, Kuwait has no other viable resource. Would the Kuwaitis accept a return to the desert? Fifty years is not a long time: in fact, it is within the lifetime of children now at school. Somewhat like Midas, the Kuwaitis cannot eat their gold; and certainly the paper money they get from the nations of the Occident is worth little. With this information about the future, the Kuwaitis have to design policies that ensure that the basic needs of life can be met a hundred years from now. In other words, trust in Allah but tie your camel.

The policies pursued so far have been largely of investment overseas.

Some investment of oil income has apparently been made by the oil-exporting countries in manufacturing and other operations of the industrialized nations. The government of Iran has, for example, recently purchased a major interest in the Krupp operations in Germany. However, such investment has its perils. This investment is probably good as long as Iran can supply the oil to run the industry that it has purchased. After it has ceased to be a supplier of this oil, however, Iran faces the possibility of its investment being confiscated by the country in which it is located, in the same way that the United States and some of its companies have been treated by other nations of the world. A good stable government, such as the Bonn government, can make commitments in good faith but it has no way of knowing what actions will be taken by a later generation, especially if that generation finds itself in a desperate condition.[2]

With such possible outcomes, the present policies could by no stretch of the imagination be considered as either good planning or good futuristic thinking. We argued in Chapter 6 that Islam favours reduction of consumption and needs and emphasizes self-sufficiency and self-reliance; and this, we believe, should be the cornerstone of all planning and futuristic thinking. The present course is diametrically opposed to this way of thinking.

Thinking seriously about the future is difficult, to say the least, when all one has is a five-year plan which says that the country will have a growth rate of 4 per cent. Development, we emphasize yet again, is a multidimensional process and requires multidimensional, culturally orientated planning. For Muslim societies, 'development planning' simply means planning for cultural development and not for 'economic growth'. As such the goal of development in Muslim societies is to steer away from occidentalism and towards Islam.

This is not just our goal; it is also our direction. It must be our permanent direction. If every change in leadership also means a change in our direction—in science policy, development strategies and cultural outlook—then we will be forever lost, taking the same narrow short-term expedient views, making the same oft-repeated mistakes. The people who do not know the direction in which they are moving only wake up when their fate has been irrevocably sealed.

It takes little imagination to foresee the outcome, in the not too distant future, of the occidentalized development strategies and policies pursued at present by the Muslim countries. It is even possible to specify, given the present trend, that:

(1) poverty will increase;
(2) unemployment among both educated and uneducated will increase;
(3) wealth will be accumulated in fewer and fewer hands;
(4) urbanization and urban population will increase;
(5) homelessness and squatters will increase:
(6) urban transport will eventually break down;
(7) technocracy and power of state-controlled technology will increase;
(8) centralization and political despotism will increase;
(9) crime and deviancy will increase—with violence becoming more rampant; and as a consequence, the cultural values of Muslim societies will be eroded away.

These have been the trends in Muslim societies over the past decades. There is no reason why they should not continue; unless, of course, positive efforts are made to reverse them.

This is the future the Muslim *Ummah* has to fight. And win.

Throughout this book we have suggested policies and strategies which could be used as weapons in this fight. Here we summarize some of these as a framework for planning, policy making and implementation and as a process orientated to future change of present situations, in contrast to occidental and trivial planning.

1. Development should be seen as neither a purely economic process or a value-free one. It means neither mass consumption nor mechanization. It is an expression of cultural dynamism characterized

by the desire of Muslims to be in 'a state of Islam'. Anything which interferes with this desire should, therefore, be seen as an impediment to development. Anything, mechanical or environmental, 'scientific' or otherwise, which interferes with their cultural values and spirituality makes a negative contribution to development, and is undesirable. On the material level, the principle of domesticity should be seen as the guiding factor for development. It is also necessary for the victims of occidentalized patterns of development to free themselves from their present structures of vulnerability.

2. Science should be considered as a cultural process; and technology as culturally subversive. When introducing conventional technology in a Muslim society, or using the results of scientific research carried out in the Occident, the inherent ideological assumptions and bias must be kept in view. A questioning approach to all scientific and technological discovery must be made. We would like to emphasise that in contrast to Neo-Apollonian rationalist thinking, Muslim modes of inquiry are not 'objective' if objectivity implies theorizing without any passionately held cultural aims or progressive linear thinking without the guidance of the spiritual values. But it is exceedingly objective if it means that every step in the process of thinking is guided not just by critically shifted evidence but also by a framework of unchanging values. The Muslims must screen, harmonize and integrate, within their cultural worldview, various forms of contemporary 'scientific' knowledge from pure to applied.

Attempts must also be made to regenerate the genuine Muslim spirit of inquiry and innovation. As such any educational policy that encourages imitation of occidentalized modes of scientific thought should be replaced by those which encourage cultural flowering, original thinking, bold use of imagination, experimentation and theory building.

3. Technological development is best achieved by concentrating on medium-size industry, and developing indigenous technologies and balanced-basic and applied-research capabilities. Rural development requires an equal amount of concentration on agriculture, animal husbandry, housing based on indigenous materials, light industries using appropriate technologies, and service organizations. Such development would need a much more sophisticated organization of material and human resources. Yet all this is within

the power of the Muslim *Ummah* and requires no outside help or foreign assistance. The images and dreams of Muslim societies must be moved away from a crowded deprivation of urban centres to green rural pastures of calm aesthetics and plenty for the basic needs of all. There must also be a shift from the present emphasis of our education systems on occidentalism, elitism and personal achievement towards cultural enlightenment and social responsibility, from urban towards rural development, from consumption towards aesthetic development.

4. The two intellectual forces in the Muslim societies, namely traditional and occidentalized, should participate in a general framework, which provides for the full flowering of those eternal values that are the real assets of the *Ummah*, and that idea which alone makes life worth living for all humanity. The traditional scholars must be accorded the status which is their due and should be involved more in general science education and in policy-making decisions.

5. The goal of 'Islamic solidarity' involves participation and cooperation between Muslim countries along their common paths to development. When planning for cooperation, there are a number of questions, in a sense quite simple and obvious, which must be asked. Should, for example, cooperation take the form of coordination or a common effort? Are there experiences in R & D which can be shared? Are there subjects of enquiry which, because of their magnitude, financial or manpower needs, require a cooperative effort? Can there be cooperation between regions over problems they face? *Et cetera*.

We think that the goals of Islamic solidarity can be achieved by cooperation sought in an effective and not in an illusionary manner. There is little to be gained by involving a *fata Morgana*, tempting for some but frightening for others. We believe that the Muslim societies should aim at what is possible in order to achieve what is (Islamically) desirable.

6. Planning in the Muslim countries must be more than a projection into the future of today's dominant trends and the current bleak outlook in many areas for chaos and disaster. It is by no means enough just to produce five-year Development Plans: these plans are nothing more than ostrich-like responses to the current situation; the plans remain dominated by current events. This is why most development planning has failed in the Muslim countries. Although

all our crises are, in reality, on the macro-level, we have concentrated all our attention and efforts on the microsystem. In the final event, it may be better to return to the beginning for action without vision, or actions that go against the very nature of the Muslim societies could prove disastrous. We suggest that planning should be at three levels:

(1) perspective planning, with at least 20-year horizons;
(2) programme planning with 5-year horizons; and
(3) budget planning with 1-year horizons.

Furthermore, planning should be continuous and adaptive. It must be continuously reviewed and recast before it does any harm to the society. It is to be remembered that

> ... even the most limited form of planning takes place within the social framework and no kind of planning can be envisaged which does not relate to man's activities. Planning may be regarded as a form of social process: in the on-going business of living we plan all sorts of activities. The way in which we plan the choices we make, the measure at our disposal for realizing our aims, and our methods of effectuation are essentially generated with the cultural patterns of our society.[3]

If we accept that our planning will be generated by the cultural patterns of Muslim society, then there is no need to base our plans on predictions that cannot really be correctly made, but only on the analysis of an unfolding situation of our *Ummah*. As such, the future is something we can determine by use of our freedom rather than something that is latent, and that takes its own course. We can make the future.

It is not fashionable in contemporary development literature to talk about visions of the future and long-range planning. Neither is it the 'done thing' to argue in favour of preservation of one's religious values and cultural patterns, or of reducing consumption and needs, or of principles of domesticity. But the Occident has become largely a world of fads and fashions. If the Muslim societies are serious about their culture, they should at least emancipate themselves from the fashions of the Occident. What matters are the problems we face, and how we operationalize practical, Islamic solutions to these problems. The policies needed to solve these problems require the Muslim societies to go against contemporary fashions to a great extent. They also require a bold effort to make an Islamic stand, as well as intellectual courage and a firm understanding of Muslim societies and the culture of Islam. Occidental trends and fashions have no role to play here.

The present study is an attempt to transcend the boundaries of the occidentalized analysis of development and contribute ideas and frameworks which take Muslim societies away from occidentalism and into a state of Islam. The future of the Muslim societies is with Islam; and without Islam they have no future.

NOTES

1. C. S. Cook, 'What Happens When the Oil Is Gone?, *Bulletin of Atomic Scientists,* June 1975, pp. 7-9; see also Z. Sardar, 'What Alternatives to Arab Oil?', *Impact International Fortnightly* 3 (13), 7 (1973).
2. Cook, 'What Happens When the Oil Is Gone?'.
3. J.-Dakin, 'An Evaluation of the "Choice" Theory of Planning', *J. Am. Inst. Plan.* 29 (19), 27 (1963).

APPENDIX 1 THE MUSLIM WORLD[1]

A. Countries Where Muslims Constitute a Majority

Country	Capital	Area km^2	Population	Muslims	%
Afghanistan	Kabul	652,015	17,900,000	17,721,000	99
Albania	Tirana	28,860	2,350,000	1,763,000	75
Algeria	Algiers	1,500,212	15,700,000	15,386,000	98
Bahrain	Manama	1,118	222,000	220,000	99
Bangladesh	Dacca	143,328	75,000,000	63,750,000	85
Cameroon	Yaounde	477,277	6,117,000	3,365,000	55
Central African Republic	Bangui	618,420	1,640,000	902,000	55
Chad	Fort Lamy	1,289,080	3,999,000	3,400,000	85
Dahomey	Porto Novo	115,154	2,909,000	1,746,000	60
Egypt	Cairo	1,005,321	35,900,000	33,387,000	93
Ethiopia	Addis Ababa	1,221,900	26,598,000	17,289,000	65
Gambia	Bathurst	10,246	384,000	327,000	85
Guinea	Conakry	245,857	4,259,000	4,047,000	95
Guinea-Bissau	Madina du Boe	36,125	810,000	567,000	70
Indonesia	Jakarta	1,491,564	131,713,000	125,127,000	95
Iran	Tehran	1,648,000	32,215,000	32,571,000	98
Iraq	Baghdad	438,446	10,164,000	9,657,000	95
Ivory Coast	Abidjan	322,500	4,515,000	2,484,000	55
Jordan	Amman	94,500	2,556,000	2,429,000	95
Kuwait	Kuwait	17,800	917,000	917,000	100
Lebanon	Beirut	8,806	3,021,000	1,722,000	57
Libya	Tripoli	1,759,500	2,178,000	2,178,000	100
Malaysia	Kuala Lumpur	286,000	11,393,000	5,925,000	52
Maldive Islands	Male	235	125,000	125,000	100
Mali	Bamako	1,239,988	5,392,000	4,853,000	90
Mauritania	Nouakchott	1,030,000	1,227,000	1,227,000	100
Morocco	Rabat	446,550	16,995,000	16,826,000	99
Niger	Niamey	1,271,896	4,355,000	3,963,000	91
Nigeria	Lagos	927,339	79,759,000	59,820,000	75
Oman	Muscat	213,200	750,000	750,000	100
Pakistan	Islamabad	1,041,375	64,892,000	62,945,000	97
Qatar	Doha	10,400	170,000	170,000	100
Saudi Arabia	Riyadh	2,158,000	8,175,000	8,175,000	100

Country	Capital	Area km^2	Population	Muslims	%
Senegal	Dakar	196,192	4,020,000	3,819,000	95
Sierra Leone	Freetown	72,605	2,769,000	1,800,000	65
Somalia	Mogadishu	702,000	3,950,000	3,950,000	100
South Yemen	Medina as-Shaab	291,200	1,560,600	1,440,000	95
Sudan	Khartoum	2,515,500	16,911,000	14,375,000	85
Syria	Damascus	186,808	6,890,000	5,994,000	87
Tanzania	Dar es Salam	943,332	14,380,000	9,347,000	65
Togo	Lome	56,600	2,120,000	1,166,000	55
Tunisia	Tunis	165,150	5,521,000	5,245,000	95
Turkey	Ankara	780,580	38,000,000	37,620,000	99
Union of Arab Emirates	Abu Dhabi	85,800	320,000	320,000	100
Upper Volta	Ouagadougou	275,259	5,514,000	3,879,000	56
Yemen	Sana'a	195,000	6,070,000	6,000,000	99
			TOTAL :	599,589,000	

Source: *The World Muslim Gazeteer* (Ummah, Karachi, 1976).

B. Areas Where Muslims Constitute a Majority

Country/Area	Capital	Area km^2	Population	Muslims	%	Political status
Azerbaijan	Baku	86,630	9,003,000	7,023,000	78	Under USSR
Brunei	Bander Seri Begawan	5,765	150,000	114,000	76	British Protectorate
Comoros Island	Moroni	2,216	290,000	286,000	95	French Overseas Territory
Eritrea	Asmara	119,000	3,000,000	2,250,000	75	Under Ethiopia
Kashmir	Srinagar	318,400	6,620,000	5,164,000	78	Under India
Kazakhstan	Alma Ata	2,766,603	12,850,000	8,738,000	68	Under USSR
Kirghizia	Frunze	199,269	2,933,000	2,699,000	92	Under USSR
Palestine	Jerusalem	26,421	3,001,400	2,612,000	87	Under Israel
Sinkiang	Urumchi	1,834,999	9,310,000	7,535,000	82	Under Peoples' Republic of China
Tajikistan	Stalinabad	140,448	2,900,000	2,842,000	98	Under USSR
Turkmenia	Ashkhabad	489,884	2,158,000	1,943,000	90	Under USSR
Uzbekistan	Tashkent	410,979	41,669,000	36,669,000	88	Under USSR
			TOTAL:	77,875,000		

C. Countries Where Muslims Constitute a Minority

Country	Area km^2	Total Population	Muslims	%
Angola	1,251,512	5,800,000	1,450,000	25.0
Argentina	2,805,569	24,290,000	486,000	2.0
Armenian S.S.R.	29,395	2,493,000	299,000	12.0
Australia	7,724,810	13,130,000	132,000	1.0
Bhutan	47,182	1,100,000	55,000	5.0
Botswana	571,519	670,000	34,000	5.0
Brazil	8,544,822	105,137,000	210,000	0.2
Bulgaria	111,270	8,620,000	1,207,000	14.0
Burma	680,568	29,560,000	2,956,000	10.0
Burundi	27,934	3,600,000	720,000	20.0
Byelorussian S.S.R.	208,400	9,003,000	540,000	6.0
Cambodia	181,734	7,000,000	70,000	1.0
Canada	10,014,680	22,130,000	100,000	0.5
Chile	744,629	10,230,000	50,000	0.05
China (less Sinkiang)	9,561,748	850,000,000	93,500,000	11.0
Congo	343,319	1,000,000	150,000	15.0
Cyprus	9,287	630,000	210,000	33.0
Guinea Baseau	28,215	300,000	75,000	25.0
Fiji Islands	18,293	550,000	60,000	11.0
Finland	338,403	4,660,000	3,000	0.06
France	552,913	52,130,000	1,043,000	2.0
Gabon	265,954	520,000	234,000	45.0
Georgian S.S.R.	69,968	4,688,000	983,000	19.0
Germany (West)	372,320	61,970,000	620,000	1.0
Ghana	238,768	9,360,000	2,808,000	30.0
Greece	132,454	8,970,000	270,000	3.0
Guyana	215,800	760,000	114,000	15.0
Hong Kong	1,016	4,160,000	42,000	1.0
Hungary	93,386	10,430,000	105,000	1.0
India	3,280,152	574,220,000	68,907,000	12.0
Italy	302,211	54,890,000	549,000	1.0
Japan	370,988	108,350,000	109,000	0.1
Kenya	584,896	12,480,000	3,682,000	29.5
Korea (South)	92,881	33,333,000	10,000	0.03
Liberia	111,800	1,660,000	498,000	30.0
Laos	237,715	3,180,000	32,000	1.0

Country	Area km^2	Total Population	Muslim	%
Lesotho	30,461	1,200,000	120,000	10.0
Malagasy Republic	592,800	6,750,000	1,350,000	20.0
Malawi	93,860	4,790,000	1,677,000	35.0
Malta	317	320,000	45,000	14.0
Mauritius	1,872	830,000	141,000	19.5
Mexico	1,980,163	54,300,000	10,000	0.02
Moldavian S.S.R.	33,831	3,572,000	107,000	3.0
Mozambique	774,100	8,820,000	2,205,000	29.0
Namibia	827,478	670,000	34,000	5.0
Nepal	141,341	12,020,000	481,000	4.0
New Zealand	269,713	2,960,000	20,000	0.6
Panama	74,758	1,570,000	50,000	3.5
Philippines	300,970	40,220,000	4,827,000	12.0
Poland	312,929	33,360,000	333,000	1.0
Portuguese Timor	19,058	640,000	128,000	20.0
Rumania	238,118	20,830,000	188,000	0.9
Russian S.F.S.R.	17,075,416	130,090,000	7,805,000	6.0
Rhodesia	390,865	5,900,000	885,000	15.0
South African Replic	1,228,133	23,720,000	474,000	2.0
Sri Lanka	65,863	13,250,000	1,195,000	9.0
Surinam	165,452	430,000	107,000	25.0
Swaziland	17,430	460,000	46,000	10.0
Taiwan	36,103	15,000,000	135,000	0.9
Thailand	520,384	39,790,000	5,571,000	14.0
Trinidad & Tobago	4,846	1,060,000	127,000	12.0
Uganda	244,350	10,810,000	3,881,000	35.9
Ukranian S.S.R.	603,319	47,136,000	5,657,000	12.0
United Kingdom			500,000	
U.S.A.	9,399,299	211,210,000	3,169,000	1.5
Vietnam	330,200	42,430,000	213,000	0.5
Yugoslavia	256,791	20,960,000	4,192,000	20.0
Zaire	2,345,409	23,835,900	2,384,000	10.0
Zambia	655,524	4,640,000	696,000	15.0
		TOTAL:	229,254,000	

D. Total Muslim Population of the World

Majority countries	599,589,000
Majority areas	77,875,000
Minority countries	229,254,000
Muslim population of the world	906,718,000

APPENDIX 2 SCIENCE AND TECHNOLOGY INFRASTRUCTURE OF SOME MUSLIM COUNTRIES[1]

Region/Country	Pop. in millions	GNP per capita[2] (USA $)	Prop. of GNP spent on R & D[3]	No. of Pub. basic scientists	Pub. basic scientists per m. population	No. of scientists & eng. per m. population[5]	No. of R & D scientists & eng. per m. population[6]
AFRICA							
Algeria	14.3	300	0.4 (1972)	28	2.0	–	17 (1972)
Tunisia	5.0	250	0.4 (1971)	13	3.0	450 (1968)	–
Egypt	33.3	210	0.5 (1971-2)	361	11.0	901 (1971-2)	198 (1968)
Sudan	15.7	120	0.6 (1971)	57	4.0	120 (1965)	–
Nigeria	55.0	120	0.5 (1970)	137	3.0	360 (1970)	38 (1970)
ASIA							
Malaysia	10.9	380	–	81	7.0	917 (1965)	32 (1966)
Philippines	36.9	210	0,25 (1968)	49	1.0	694 (1968)	–
Pakistan (& Bangladesh)	130.2	100	0.3 (1963-4)	99	0.8	54 (1968)	⋮
Afghanistan	14.0	80	–	8	0.1	345 (1966)	–
EUROPE							
Turkey	35.2	310	0.75 (1967)	88	30.0	166 (1967)	5 (1967)
MIDDLE EAST							
Lebanon	2.7	590	0.3 (1967)	67	25.0	1,901 (1969)	84 (1967)
Saudi Arabia	7.4	440	0.06	7	1.0	–	–
Iran	28.7	380	0.5 (1970)	92	3.0	3,454 (1970)	104 (1970)
Iraq	9.7	320	0.06	41	4.0	–	–
Syria	6.0	290	0.06	3	0.3	–	–

To safeguard the interests of the developing countries from inappro-
priate technologies, the Working Party of Experts in the Pugwash
Conference on Science and World Affairs has produced the following
Draft Code of Conduct on Transfer of Technology.[1]

Objectives and Principles

(1)   To establish general equitable rules of behaviour in the international technology markets taking into consideration particularly the
justified needs of developing countries and legitimate rights and obligations of technology producers and suppliers and technology recipients.

(2)   To make clear the distinction between proprietary technology
and freely available technology and reflect this distinction in terms and
conditions of technological transactions.

(3)   To foster the expansion of international trade in proprietary
technology on terms mutually beneficial to suppliers and recipients by
eliminating restrictive technology trade practices and regulating monopolistic rights accruing to some proprietary technology owners for the
purpose of assuring the strengthening of the negotiating power of
developing countries.

(4)   To ensure fair pricing of technology trade transactions, by
assessing all direct and indirect costs to the recipients and profits to the
suppliers, and taking into consideration, *inter alia*, the duration of a
contract and the dynamics of technological progress.

(5)   To introduce as a minimum the most-favoured-licensee clause
in the international technology trade transactions involving developing
countries.

(6)   To expand free flow of non-proprietary technology on a non-
discriminatory basis and through appropriate channels and mechanisms
to suit the requirements of developing countries.

(7)   To ensure the responsibility of suppliers and recipients to
adapt technological trade transactions and flows of freely available
technology to factor proportions of the countries with different
development levels and to their local development needs and absorptive
capacity.

(8)   To increase the contribution of technology, under specially favourable conditions, for the solution of pressing social problems in developed and developing countries.

(9)   To ensure that technological transactions entail the strengthening of local technological capability of developing countries, which would permit their indigenous technological dependence upon the outside world and assuring their increasing participation in world technology production and trade . . .

Guarantees

The supplier shall guarantee that:

(1)   the technology acquired is in itself suitable for manufacture of products covered by the agreement;

(2)   the content of the technology transferred is in itself full and complete for the purposes of the agreement;

(3)   the technology obtained will in itself be capable of achieving a predetermined level of production under the conditions specified in the agreement;

(4)   national personnel shall be adequately trained in the operation of the technology to be acquired and in the management of the enterprises;

(5)   the technology is the most adequate to meet the particular technological requirements of the recipient given the supplier's technological capabilities;

(6)   the recipient shall be informed and supplied with all improvements on the techniques in question during the lifetime of the agreement;

(7)   where the recipient of the technology has no other technological alternative than acquiring capital goods, intermediate inputs and/or raw materials from or selling his output to the technology supplier or any source designated by him, the prices of the articles shall be consonant with current international price levels;

(8)   for a certain period of time the supplier shall guarantee to provide spare parts, components, and servicing of the technology without additional charges for maintaining this guarantee;

(9)   all transfer of technology arrangements should include a provision by which if the licensor grants more favourable terms to a second licensee these terms will be automatically extended to the first licensee.

The recipient shall guarantee that:

(1) the acquired technology will be used as specified in the contract;

(2) all legitimate payments as specified in the contract shall be made to the technology suppliers;

(3) the technical secrets as defined in the contract shall be honoured;

(4) the quality standards of the products specified in the contract will be reached and maintained where the contract includes the use of the supplier's trade mark, trade name, or similar identification of good will;

(5) the socio-economic conditions and needs of the country of the recipient have been taken into account while entering into a technology transfer agreement . . .

Measures According Special Treatment to Developing Countries

Transfers of technology to the developing countries shall include forms of preferential treatment designed to take account of the weaker position of their enterprises in the technological, financial or managerial field. Such measures shall include, *inter alia*

(1) phasing out of down payments, or including such payments as part of royalties on production, on a soft basis;

(2) scaling down of the charges for technology in proportion to the size of the recipient's market;

(3) untying of credits for the purchase, from the most competitive source, of capital goods, spare parts and intermediate components;

(4) rebates on imports of raw materials, equipment and components for licensed production;

(5) development of local technological capability and research and development by technology suppliers with affiliated companies in developing countries;

(6) development of the research and development and technological capability of the recipient firm;

(7) adapting the technology to be transferred to make it appropriate to conditions and factor endowment of the recipient country;

(8) transferring to the recipient firm non-proprietary technology which the supplier may possess in the field of activity of the recipient; sub-licensing rights under special concessional terms . . .

NOTES

Appendix 1
1. Source: *The World Muslim Gazeteer* (Umma, Karachi, 1976).

Appendix 2
1. Figures are for 1965-70.
2. From *World Bank Atlas* (International Bank for Reconstruction and Development, 1972). The data are for 1975.
3. The available data consists largely of estimates. The year to which the data pertain is given in brackets, unless the figure is an estimate for the entire 1965-70 period. Sources: *UNESCO Statistical Year Book* (Paris, 1975).
4. From *International Directory of Research and Development Scientists 1969.*
5. Estimates only. Statistics include all scientific manpower in a country including those scientists who are not working specifically in R & D. Social scientists are not included.
6. Includes social scientists.

Appendix 3
1. See the report of the international group on transfer of technology, UN Trade and Development Board, JD/B/520 of 6 August 1974.

BIBLIOGRAPHY

Abdoulaye L.Y. *Les Masses africaines et l'actuelle condition humaine.* Edition Présence Africaine, Paris, 1956.

Adams, Walter (ed.). *The Brain Drain.* Macmillan, New York, 1968.

Ahmad, K. 'Are We Asking the Right Questions?' *Impact International Fortnightly* 4 (10), 1-2 (1974).

————. 'Economic Development in Islamic Framework.' Paper presented at the International Conference on Islamic Economics, Mecca, 5-11 April 1975.

————. (ed.) *Islam: Its Meaning and Message.* Islamic Council of Europe, London, 1976.

Ahmad, M. 'The Organisation of Science: Orient and Occident.' *Impact of Science on Society* 26, 2/3 (1976).

Ahmad, S. M. *Economics of Islam: A Comparative Study.* Ashraf, Lahore, 1972.

Alatas, S. H. *The Sociology of Corruption.* Donald Moore, Singapore, 1968.

Allen, G. R. and Smethurst, R. G. *Impact of Food Aid on Donor and Other Food Exporting Countries.* FAO, Unipub, 1965.

Allison, A. (ed.). *Population Control.* Penguin, London, 1970.

Allison, Samuel K. 'Physics in Egypt: A New Type of Lend-Lease.' *Bulletin of the Atomic Scientists* 16, 317 (1960).

Almond, G. A. and Verba, S. *The Civic Culture.* Princeton University Press 1963.

Almond, G. A. and Coleman, J. S. (eds.). *The Politics of Developing Areas.* University Press, 1970.

Alpert, P. *Partnership or Confrontation? Poor Lands and Rich.* The Free Press, New York, 1973.

Anawati, G. 'Moslem Science: A Theosophic-Historical View.' *Impact of Science on Society* 26, 2/3 (1976).

Anderson, C. W. *et al. Issues of Political Development.* New Jersey, 1967.

Ansari, J. 'Economic Planning: The Central Task.'*Impact International Fortnightly* 1 (12), 6 (1971).;

————.'Must We Pay an Exploitative Price?' *Impact International*

Fortnightly 1 (6), 5-6 (1971).

————. 'The New Slave Trade.' *Impact International Fortnightly* 1 (4), 6-7 (1974).

————. 'Of People, Not of Things.' *Impact International Fortnightly* 1 (1), 5 (1971).

Ashraf, S. A. 'Islam and Modern Scientific Attitude.' *Impact International Fortnightly* 4 (2), 6-7 (1974).

Austin, O. 'Islam in the Modern World.' *The Muslim* 12 (3), 58-60 (Feb./Mar. 1975).

Aziz, K. K. *The Making of Pakistan.* Chatto and Windus, London, 1967.

Aziz, S. 'The World Food Problem.' *New Internationalist* 29 (July 1975).

Azzam, A. R. *The Eternal Message of Muhammad.* Davin Adsir Co., New York, 1964.

Badion, S. *Les dirigeants africains face à leur peuple.* F. Maspero, Paris, 1964.

Bailey, F. *Tribes, Castes and Nations.* Manchester, 1966.

Bammate,H. *Visages de l'Islam.* Paris, 1955.

Baldwin, G., 'Brain Drain or Overflow?' *Foreign Affairs* 48 358 (1970).

Ball, N. 'The Myth of the Natural Disaster.' *Ecologist* 5 (10) 368-71 (1975).

Banage, W. B. 'The Development of Science in East Agrica.' *Scientific World* 11, 15 (1967).

Banjo, A. 'Pros and Cons of Intermediate Technology: Technology and Development.' *Unitar News* 6 (4), 11-12 (1974).

Basalla, George. 'The Spread of Western Science.' *Science* 156, 611 (1967).

Bauer, P. T. 'Foreign Aid, Forever?' *Encounter* 42 (3), 15-29 (March 1974).

Baun, R. J. 'A Philosophical/Historical Perspective on Contemporary Concerns and Trends in the Area of Science and Values.' *Newsletter of the Program of Public Conception of Science*, Harvard University, 1974.

Bawany, E. A. 'Revolutionary Strategy for National Development.' *Muslim News International*, Karachi, 1970.

Beckerman, W., *In Defence of Economic Growth.* Cape, London, 1974.

Bell, D. *The End of Ideology.* Free Press of Glencoe, New York, 1960.

Bernal, J. D. *Science in History.* Penguin, London,1970.

Benjamin, A. C. *Science, Technology and Human Values.* U. Mo. P., 1965.

Bernard, H. R. and Pelto, P. J. *Technology and Cultural Change.* Macmillan, London, 1972.

Bhabha, H. J. 'Science and the Problems of Development.' *Science* 151, 541 (1966).

Bhathal, R. S. 'Science and Technical Education in Asia.' *New Scientist and Science Journal* 729 (June 1971).

————. 'Science Policy in the Developing Nations.' *Nature* 232, 227 (1971).

————. 'Science, Religion, and Society.' *Journal of the Political Science Society* (Singapore) (November 1971), p. 8.

Binder, L. *The Ideological Revolution in the Middle East.* Wiley, New York, 1964.

Binder, L. *et al. Crisis and Sequences in Political Development.* Princeton University Press, 1965.

Blackett, P. M. S. 'Planning for Science and Technology in Emerging Countries.' *New Scientist* 17, 345 (1963).

Blackledge, J. P. 'The University as an Adaptor of Technology in a Developing Country.' Denver Research Institute, University of Denver, Denver, Colorado 1972 (unpublished).

de Bono, E. *Technology Today.* R. & K. Paul, London, 1971.

Bose, T. C. *The Superpowers and the Middle East.* Asia Publishing House, 1972.

Boulding, K. E. and Mukerjee, T. (eds.). *Economic Imperialism: A Book of Readings.* University of Michigan Press, 1972.

Braibenti, R. and Spengler, J. J. (eds.). *Traditions, Values and Socio-Economic Development.* Duke University Press, 1961.

Bronowski, J. *Science and Human Values.* Harper Torchbooks, New York, 1965.

Brooks, H. 'Can Science be Re-directed?' Paper circulated at Colloque organized by Conservatoire National des Arts et Métiers (CNAM), on *Peut-on rediriger la science?*, Paris, 4-6 December, 1965.

————. H. Letter to Minerva, *Minerva* 10, 327-8 (April 1972).

Brown, L. 'The Green Revolution', in Ward, B. *et al.* (eds.). *The Widening Gap.* Columbia University Press, 1971.

Burke, J. G. (ed.). *The New Technology and Human Values.* Wansworth, Belmont, 1966.

Burkhardt, G. 'Science Education in Africa.' *Bulletin of the Atomic Scientists* 22, 46 (1966).

Cardoso, F. H. *Ideologias de la Burguesia Industrial et Sociedades Dependientes (Argentina y Brazil).* Siglo XXI Editores, Mexico, 1971.

Carney, D. *Government and Economy in British West Africa.* Yale University Press, 1972.

Carol, Sir Olaf. *Soviet Empire.* Macmillan, London, 1967.

Celasun, M., 'Technological Advance as a Factor in the Turkish Development Planning: Some Observations and Suggestions.' *Management of Research and Development* (Proceedings of the Istanbul Seminar on, Research and Development Management), OECD, Paris, 1972, pp. 11-32.

Cervantes, J. L. C. *Superexplotacion, Dependencia y Desarrollo.* Editorial Nuestro Tiempo, Mexico, 1970.

Chapra, M. U. *The Economic System of Islam — A Discussion of its Goal and Nature.* Islamic Cultural Centre, London, 1970.

de Chardin, P.T. *The Future of Man,* Fontana, London, 1955.

————. *Man's Place in Nature.* Fontana, London, 1966.

————. *The Phenomenon of Man.* Fontana, London, 1955.

Chibwe, E. C. *Arab Dollars for Africa.* Croom Helm, London, 1976.

CIBA Foundation. *Civilization and Science — In Conflict or Collaboration?* Elsevier, Amsterdam, 1972.

Clarke, R. *The Great Experiment: Science and Technology in the Second United Nations Decade.* UN Centre for Economic and Social Information, New York, 1971.

————. 'The Pressing Need for Alternative Technology. ' *Impact of Science on Society* 23 (4), (1973).

Coats, J. 'Technology Assessment: The Benefits, The Costs, The Consequences.' *The Futurist* 5 (6) (December 1971).

Coile, R. C. 'Educational Planning in Developing Countries.' Paper presented at the Société Française de Recherche Operationelle Symposium on 'The Possibilities of Operational Research in Developing Countries', Paris 26-8 June 1963.

Commoner, B. *The Closing Circle.* Cape, London, 1972.

————. *Science and Survival.* Ballantine, New York, 1963.

Cook, C. S. 'What Happens When The Oil Is Gone?' *Bulletin of Atomic Scientist,* June 1975.

Cooper, C. (ed.). 'Science and Underdeveloped Countries', in *Problems of Science Policy.* OECD, Paris, 1967.

————. (ed.). *Science, Technology and Development.* Frank Cass, London, 1973.

Copeland, M. *The Game of Nations.* Allen & Unwin, London, 1969.

Crane, D. 'An Inter-Organisational Approach to the Development of Indigenous Technological Capabilities: Some Reflections on the Literature.' OECD Occasional Paper No. 3, CD/TI(74)31. 1974.

————. *Invisible Colleges*. University of Chicago Press, 1972.

Dainton, Sir Fredrick. *The Future of the Research System*. Comm. 4814 HMSO, London, 1972.

Dakin, J. 'An Evaluation of the Choice Theory of Planning.' *J. Am. Int. Plan.* 24, 19-27 (1963).

Dar, B. A. *Quranic Ethics*. Institute of Islamic Culture, Lahore, 1960.

Dart, Francis and Pradhan, Panna Lal. 'Cross-Cultural Teaching of Science.' *Science* 155, 649 (1967).

————. 'The Cultural Context of Science Teaching.' *Search* 4, 48 (1973).

————. 'Readiness in Abstraction.' In *Education in Developing Countries of the Commonwealth*. Commonwealth Secretariat, London, 1973.

————. 'Science and the Worldview.' *Physics Today* 25, 48 (1972).

Davenport, W. H. *The One Culture*. Pergamon, London, 1970.

Dedijer, Stevan. 'Scientific Research and Development: A Comparative Study.' *Nature* 187, 458 (1960).

————. 'Underdeveloped Science in Underdeveloped Countries.' *Minerva* 2 (1), 61 (1963). Reprinted in Shils, Edward. *Criteria for Scientific Development: Public Policy and National Goals*. MIT Press, Cambridge, Mass., 1968.

Dickson, D. *Alternative Technology and Politics of Technical Change*. Fontana/Collins, London, 1974.

Deutsch, K. W. *Nationalism and Social Communication*. MIT Press, Cambridge, Mass., 1953.

————. *Politics and Government*. Houghton Mifflin, Boston, 1970.

Djerassi, Carl. 'A High Priority: Research Centers in Developing Nations.' *Bulletin of the Atomic Scientists* 24, 22 (Jan. 1968).

Dorozyinski, A. 'Eden With Oil Wells.' *Nature* 257, 78-87 (1975).

Dubos, R. *Reason Awake: Science for Man*. Columbia University Press, 1971.

Dumont, R. *L'Afrique Noire est mal partie*. Editions du Seuil, Paris, 1962.

Duncanson, D. J. *Government and Revolution in Vietnam*, OUP, 1974.

Edwards, M. *Raj*. Pan Books, London, 1969.

Ehrlich, P. R. and Ehrlich, A. *Population, Resources, Environment. Issues In Human Ecology*. Freeman, 1970.

Elliot, C. 'The Poorest of the Poor.' *New Internationalist*, 9 November 1973.

Al-Emri, B. O. 'Modernization and the Islamic World. A Theoretical Analysis.' Paper presented at the International Conference on Islamic Economics, Mecca, 5-11 April, 1975.

Epstein, A. L. *Politics in an Urban African Community.* Manchester University Press, 1958.

Fairbairn,G. *Revolutionary Guerrilla Warfare. The Countryside Version.* Pelican, London, 1975.

Fanon, F. *Wretched of the Earth.* London, 1965.

Faruki, K. A. 'Early and Pre-Modern Approaches to Muslim Unity.' *Impact International Fortnightly* 3 (8), 6-7 (1973).

————. 'Pan-Islamicisation and Islamic Universalism.' *Impact International Fortnightly* 3 (9), 10-11 (1973).

Faruqi, M. H. 'On the Basis of Culture in Islam.' *The Muslim* 8 (4), 81-4 (Jan./Feb. 1971).

————. 'Revolution or Renovation?' *Impact International Fortnightly* 2 (14), 8-9 (1972).

Federov, E.K. 'Some Problems Relating to Developing Countries.' *Impact* 13 (4), 273 (1963).

Fei, J. C. H. and Goston, R. *Technology Transfer, Employment and Development.* Yale Economic Growth Centre, Discussion Paper No. 17, undated.

Feyerbend, P. *Against Method.* NCB, London, 1975.

Fleming, Launcelot. 'Living With Progress.' *New Scientist* 36, 166 (1967).

Freeman, C. 'Measurement of Output of Research and Experimental Development — A Review Paper.' UNESCO Statistical Reports and Studies No. 16, Paris, 1969.

Freeman, C. and Young, A. *The Research and Development Effort of Western Europe, North America and Soviet Union.* OECD, Paris, 1966.

Fromm, E. *The Revolution of Hope: Towards A Humanized Technology.* New York, 1968.

Fry, M. J. *Development Planning in Turkey.* Brill, Leiden, 1971.

Fuller, J. F. C. *The Conduct of War.* Allen & Unwin, London, 1961.

Furnivall, J. S. *Colonial Policy and Practice.* CUP, 1948.

Furtado, C. *Development and Underdevelopment.* University of California Press, 1964.

Geertz, C. C. (ed.). *Old Societies and New States.* Free Press of Glencoe, New York, 1963.

Geliner, O. *The Enterprise Ethic.* London, 1967 (ch. VIII: 'International

Relations and the Competitive Economy. The Case of Underdeveloped Countries.')

Ghana Council for Scientific and Industrial Research. *Scientific Research in Ghana.* SCIR, Accra, 1971.

———. *Workshop on the Role of the Council for Scientific and Industrial Research in Determining Science Policy and Research Priorities.* Accra, CSIR, 1973.

Al-Ghazzali. *The Book of Knowledge.* Translated by Nabih Amin Faris. Ashraf, Lahore, 1962.

Gill, R. T. *Economic Development — Past and Present.* Prentice-Hall, 1973.

Ginzberg, E. (ed.). *Technology and Social Change.* Columbia University Press, 1961.

Golay, F. H. *The Philippines: Public Policy and National Economic Development.* Cornell UP, Ithaca, New York, 1961.

Goulet, G. *The Cruel Choice.* New York, 1971.

Gowon, Yakubu. 'Science Technology and Nigerian Development.' *Impact* 22 (1/2), 55 (1972).

Graves, M. 'Developing Country Ltd.' *New Internationalist* 11 (2), 21 (Jan. 1974).

Gruber, Ruth (ed.), *Science and the New National.* New York, Basic Books, 1961.

Gudman, P. 'Can Technology be Humane?' *New York Review of Books* 13, 27-34 (Nov. 1969).

Guest, I. 'Is the World Working?' *New Internationalist* 39 (May 1976).

Haider, S. J. 'Science-Technology Libraries in Pakistan.' *Special Libraries* 65 (10/11) 474-8 (1974).

Hakim K. A. *Islam and Communism.* Institute of Islamic Culture, Lahore, 1969.

El-Haljowy, M. 'Recent Trends and Developments in Industrial Research in the Developing Countries of Middle East.' Paper presented at the UN Industrial Development Organization Seminar for the Stimulation of Industrial Research in Developing Countries, Singapore, 1972.

Halpern, M. *The Politics of Social Change in Middle East and North Africa.* Princeton University Press, 1963.

Hamidullah, M. *Introduction to Islam.* Centre Cultural Islamique, Paris, 1959.

Haq, M. U. 'Wasted Investment in Scientific Research.' In Moorehouse, W. 'Confronting a Four-Dimensional Problem: Science, Technology,

Society and Tradition in India and Pakistan.' *Technology and Culture* 8, 363 (1967).

Hayami, Y. and Rutton, V. W. *Agriculture Development. An International Perspective*. Johns Hopkins Press, Baltimore and London, 1971.

Hayter, T. *Aid as Imperialism*. Penguin, London, 1971.

Heady, F. *Bureaucracies in Developing Countries: Internal Roles and External Assistances*. Bloomington, 1966.

Herrera, A. 'Social Determinants of Science Policy in Latin America.' *Journal of Development Studies* 9, 19-37 (October 1972).

Hertman, P. *Society and Assessment of Technology*. OECD, Paris, 1973.

Hetzler, S. *Technological Growth and Social Change*. Praeger, New York, 1969.

Hirshman, A. *The Strategy of Economic Development*. Yale University Press, 1958.

Holton, G. 'On Being Caught Between Dionysius and Apollonius.' *Daedalus*, Summer 1974.

Hunke, S. *Allahs Sonne über dem Abendland: unser arabisches Erbe*. Stuttgart, 1960.

Huntington, S. P. *Political Order in Changing Societies*. Yale University Press, 1968.

Huxley, A. *Brave New World*. Chatto and Windus, London, 1932.
———. *Brave New World Revisited*. Harper & Row, New York, 1965.

Idris, J. S. 'Rationality, Science and Unbelief.' *Impact International Fortnightly* 4 (22), 1-2 (1074).

Illich, I. *Celebration of Awareness*. Calder & Boyars, London, 1971.
———. *Energy and Equity*. Calder & Boyars, London, 1971.
———. *Medical Nemesis*. Calder & Boyars, London, 1975.
———. *Tools for Conviviality*. Calder & Boyars, London, 1975.

Impact. 'From Poverty to Pauperism.' 1 (24), 1-2 (1972).
———. 'The Problem of Arab Billions.' 4 (1), 1 (1974).

Indonesia, Government of. *The First Five-Year Development Plan* (1969/70-1973/74). 6 vols. See particularly vol. 2c. Government of Indonesia, Djakarta, 1969.
———. *Indonesian Institute of Science (Lembaba Ilma Pengetahuan Indonesia)*. LIPI, Djakarta, 1971.

Interdoc. *Guerrilla Warfare in South East Asia*. Interdoc, 1973.

Irving, T. B. 'The Muslim World: Tasks and Perspectives.' *Impact Inter-*

national Fortnightly 5 (6), 8-10 (1975).

Janowitz, M. *The Military in the Political Development of New Nations.* University of Chicago Press, 1964.

Jenkins, R. *Exploitation.* Paladin, London, 1971.

Johnson, H. G. *The Role of Military in Underdeveloped Countries.* Princeton University Press, 1962.

Johnson, S. *The Green Revolution.* Hamish Hamilton, London, 1972.

Johnston, B. F. 'Agriculture and Structural Transformation in Developing Countries. A Survey of Research.' *Journal of Economic Literature* 8 (2), 369-403.

Jolly, R. 'The Aid Relationship: Reflections on the Pearson Report.' In Ward, B. *et al.* (ed.). *The Widening Gap: Development in the 1970's.* Columbia University Press, New York, 1971.

————. *et al.* (eds.). *Third World Employment.* Penguin, London, 1973.

Jones, G. *et al. Planning Development and Change. A Bibliography of Development and Change.* University of Hawaii Press, 1971.

————. *Role of Science and Technology in Developing Countries.* OUP, 1971.

Kadir, N. A. 'A Unified Research Council for the Arab States.' *Impact of Science on Society* 26 (2/3) (1976).

Kettani, A. 'The Scientific Heritage of Islam.' *Impact of Science on Society* 26 (2/3) (1976).

Khan, Mohammed Ali. 'Education and Research in Pakistan.' *Impact* 19 (1), 85 (1969).

Khurshid, A. *et al. Fact Sheets on Libraries in Islamic Countries.* University of Karachi, 1974.

Khusro, A. M. 'Economic Laws of Majority-Minority Relations.' Paper presented at Muslim Institute Seminar, London, 21 August 1976.

Kirby, E. S. *Economic Development in East Asia*, Allen & Unwin, London, 1967.

Kong, E. 'Agriculture vs Industrial Development in LDCs' *Intereconomics* Hamburg 6, 177-80 (1972).

Lambs, T. 'Relevance of Western Education to Developing Countries.' *The Teilhard Review* 11 C1 2-9 (1976).

Leiserson, A. 'The Politics of Science: Science Politics, Science Policy; Policy Science—The Whole Thing.' *Policy* 6 (1) (Autumn 1973).

Lengelle, M. 'Labour Productivity in Agriculture and the Balance Between the Three Main Sectors of the Economy in Developing Countries.'

Revue Européenne des Sciences Sociales 10 (26), 223-37 (1972).

Lerner, D. *The Passing of Traditional Society*. New York, 1958.

Lewis, A. W. 'Needs of New States: Science, Men and Money.' *Bulletin of the Atomic Scientists* 17, 43 (February 1941).

————. 'Education for Scientific Professions in the Poor Countries.' *Daedalus* 91, 310 (1962).

Leys, C. *Politics and Change in Developing Countries: Studies in Theory and Practice of Development*. CUP, 1961.

Lim, D. 'The Role of University in Development Planning in Malaysia.' *Minerva* 12, 18-32 (1974).

Lipton, M. *Why Poor People Stay Poor*. Temple Smith, London, 1976.

Little, I., Scitowski, T., and Scott M. *Industry and Trade in Some Developing Countries*. OUP, 1970.

Malaysian Ministry of Education. *Islamic Education in Malaysia*. Kuala Lumpur 1976

Malecki, I. 'Some Problems Concerning Organisation of Scientific Research in the Developing Countries' *Impact* 13, 181 (1963).

Mannon, M. A. *Islamic Economic Theory and Practice*. Ashraf, Lahore, 1976.

Manod, J. *Chance and Necessity*. Fontana, London, 1974.

Marcuse, H. *One Dimensional Man*. R. & K. Paul, London, 1964.

Marsden, K. 'Progressive Technologies for Developing Countries.' In Galensen, W. (ed.). *Essays on Employment*. ILO, Geneva, 1971.

Maslow, A. H. *Motivation and Personality*. Harper and Row, New York, 1954.

Mandudi, A. A. *Islamic Way of Life*. Islamic Publications, Lahore, 1965.

————. *Economic Problems of Man and their Islamic Solution*. Islamic Publications, Lahore, 1969.

Mead, M. and Metroux, R. (eds.). *The Study of Culture: At a Distance*. University of Chicago Press, 1949.

Mehdi, S. 'Reformism: A Study in Method.' *Suara Al-Islam* 2 (6), 15-30 (June 1976).

von der Mehden, F. *Religion and Nationalism in South East Asia*. University of Wisconsin Press, Madison, 1963.

Mendelsohn, E. 'A Human Reconstruction of Science.' *Boston University Journal* (Spring 1973).

————. 'Should Science Survive its Success?' In Cohen, R. S. *et al.* (eds.). *For Dirk Struik*. D. Reidel, pp. 373-89.

Merhav, M. *Technological Dependence, Monopoly and Growth*, Pergamon, London.1968.

Mesthene, E. G. *Technological Change.* Mentor, New York, 1970.

Metcalf, D. *The Economics of Agriculture.* Penguin, London, 1969.

Mikdashi, Z. *The Community of Oil Exporting Countries. A Study in Governmental Co-operation.* Allen & Unwin, 1972.

————. and Shlavim, A. 'OPEC and the Politics of Oil.' In *Organisation in World Politics Year Book 1975.* Croom Helm, London, 1976.

Millikan, M. P. and Hopgood, D. *No Easy Harvest.* Little Brown, Boston, 1967.

Milton, J. *Careless Technology: Ecology and International Development.* Tom Stacey, London, 1972.

Mingay, E. *The Agricultural Revolution.* Allen & Unwin, London, 1967.

Mintz, J. S. *Mohammed, Marx and Marhaen.* Praeger, New York, 1965.

Miraboglu, M. *Aspects of the Turkish Brain Drain.* UNESCO document SC. 72/CONF 3/3 Annex 24 (2), Paris, 1972.

Miscond, C. A. *Tunisia, The Politics of Modernisation.* Praeger, New York, 1964.

Mishan, E. J. *Technology and Growth: The Price We Pay.* Praeger, New York, 1970.

Mitchison, Naomi. 'Science in Egypt.' *The New Scientist* 7, 1073 (1960).

Metroff, I. *The Subjective Side of Science.* Elzevier, Amsterdam, 1974.

Montgomery, J. D. and Siffin, W. J. *Approaches to Development: Politics Adminstration and Change.* New York, 1966.

Moore, B. *The Social Origins of Dictatorship and Democracy.* Pelican, London, 1974.

Moravcsik, M. J. 'A Case of a Scientist in Pakistan's Social Order—a Different View.' *The Nucleus* 2-1, 19 (1965).

————. 'Fundamental Research in Underdeveloped Countries.' *Physics Today* 17 (1), 21 (1964).

————. 'Science and Technology in National Development Plans: Some Case Studies.' US Agency for International Development, Washington D.C., 1973, unpublished.

————. *Science Development: The Building of Science in Less Developed Countries,* International Development Research Center, Indiana University, Indiana, 1975.

Moorehouse, W. 'Confronting a Four-Dimensional Problem: Science, Technology, Society and Tradition in India and Pakistan.' *Technology and Culture* 8, 363 (1967).

Moorehouse, W. (ed.). *Science and the Human Condition in India and Pakistan.* Rockefeller University Press, New York, 1968.

Mosharrafa Pasha, A. M. 'The Egyptian Academy of Sciences.' *Nature*

157, 573 (1946).

Moss, R. *The Santiago Model, Conflict Studies 31 and 32*. Institute for the Study of Conflict, London, Jan. 1973.

————. *Chile's Marxist Experiment*, Halsted Press, New York, 1974.

Mueller, H. N. and Steffons, H. J. (eds.), *Science, Technology and Culture*. AMS Press, 1974.

Mukerjee, P. K. *Underdevelopment, Educational Policy and Planning*. Asia Publishing House, 1968.

Muller, J. H. *The Children of Frankenstein*. Indiana University Press, Bloomington, 1970.

Mullick, M. A. H. 'The Aid Euphemism.' *Impact International Fortnightly* 3 (10), 1 (1973).

————. 'Backlog and Development in the Muslim World.' *Impact International Fortnightly* 5 (2), 9-10 (1975).

————. 'Pakistan: Independence or Aid-Dependence.' *Impact International Fortnightly* 4 (17), 10 (1974).

————. 'The Problem of Development and Unemployment.' *Impact International Fortnightly* 1 (17), 6-8 (1971).

————. 'Underdevelopment of East Pakistan: Getting Out of An Unfortunate Legacy.' *Impact International Fortnightly* 1 (12), 7-9 (1971).

Muriel, A. 'Brain Drain in the Philippines.' *Bulletin of the Atomic Scientists* 26, 38 (1970).

The Muslim. 'Pakistan Ulema Lay Down Outlines for Economic Reforms.' 7 (9), 202-4 (June 1970).

Muslim Institute. *Draft Prospectus*. Open Press, Slough, 1971.

————. 'The Fourteenth Century of Hijra' A Memorandum submitted to Islamic Secretariat, Jeddah, 1975.

Myatt, G. 'Scientific and Technical Information on Indonesia—Problems and Prospects.' *BLL Review* 1 (2), 52-6 (1973).

Myrdal, G. *Asian Drama. An Enquiry Into the Poverty of Nations*. New York, 1968.

Nader, C. and Zahlan, A. B. *Science and Technology in Developing Countries*. CUP, 1969.

El-Naggar, F. 'The Methodology of Islamic Economics: A Systems Theory Model.' Paper presented at the International Conference on Islamic Economics, Mecca, 5-11 April 1975.

Nasr, H. *The Encounter of Man and Nature*. Allen & Unwin, London, 1968.

————. *Islamic Science*. World of Islam Festival Trust, London,

1976.

Nasser, G. A. *The Philosophy of the Revolution.* Smith Keynes and Marshall, New York, 1959.

Nature 1964. 'Science Planning, Development and Co-operation in the Countries of the Middle East and North Africa: 189, 362 (1964).

————. 'Scientists Gather in Riyadh' 260, 663 (1976).

Nayudamma, Y. 'Promoting the Industrial Application of Research in an Underdeveloped Country.' *Minerva* 5, 323 (1967).

Nettle, J. P. 'Strategies in the Study of Political Development.' In Leys, C. (ed.) *Politics and Change in Developing Countries.* CUP, 1969.

Newmark, P. 'Iranian Transplant.' *Nature* 261, 358-9 (1976).

Nigerian Council for Science and Technology *First Annual Report.* Lagos, NCST, 1970.

Nigeria, Government of. *Second National Development Plan 1970-4.* Lagos, NCST, 1970.

Niromand, B. *Iran: The New Imperialism in Action.* Modern Reader Paperbacks, New York and London, 1969.

Northrop, E. P. 'Improving Science Education in Turkey.' *Yeni Ortaogretim* [New Secondary Education] no. 5 (Feb. 1965).

Nutty, L. *The Green Revolution in West Pakistan: Implications of Technological Change.* Praeger, New York, 1972.

Odhiambo, T. 'East Africa: Science for Development.' *Science* 158, 876 (1961).

Ohyar, O. 'The University and Regional Development.' In C. Nader and A. B. Zahlan, *Science and Technology in Developing Countries.* CUP, 1969.

————. 'Universities in Turkey.' *Minerva* 6, 213 (1968).

Oldham, C. H. G. 'Science and Education.' *Bulletin of the Atomic Scientists* 22, 40 (June 1966).

Oppenheimer, M. *Urban Guerrilla.* Pelican, London, 1974.

Organization for Economic Cooperation and Development. 'Relating Science and Technology to Economic Development—a Five Country Experiment.' *OECD Observer* 15, 8 (1965).

————. 'Scientific Research and Technology in Relation to the Economic Development of Turkey. Directorate of Scientific Affairs: Pilot Teams' project on Science and Economic Development.' OECD, 1969.

Organski, A. F. K. *Stages of Political Development.* Knopf, New York, 1965.

Al-Otaiba, M. S. *OPEC and the Petroleum Industry.* Croom Helm,

London, 1975.

Otieno, N. C. 'Today's Schools Prepare Tomorrow's African Scientists.' *UNESCO Courier* 20 (June 1967).

Ozinona, A. K. 'Patterns of Scientific Development in Turkey, 1933-66.' In C. Nader and A. B. Zahlan.

von Paczensky, G. *Die Weifen Kommen: Die Wahre Geschichte des Kolonialismus.* Hoffman u Campe, Hamburg, 1971.

Paradas, K. *Technological Choice Under Development Planning.* International Publishing Service, 1963.

Pearson, L. *Partners in Development.* Praeger, New York, 1969.

Pennoch, J. R. (eds.). *Self-Government in Modernizing Nations.* Prentice-Hall, New Jersey, 1964.

Philippines, National Science Development Board. *US Workshop in Industrial Research.* Part II, *Working Papers.* Baguio City, 1969.

Pickthall, M. M. *The Cultural Side of Islam.* Ashraf, Lahore, 1961.

Platt, J. 'Universities as Nerve Centres of Society.' In *Future as an Academic Discipline.* CIBA Foundation Symposium, 36 North Holland, Amsterdam, 1965.

Press, L. 'Scientific Research Institution in Asia.' *Impact of Science on Society* 19 (1), 25-51 (1969).

Price, D. de Solla., 'Nations Can Publish or Perish.' *International Science and Technology,* October 1967.

————. *The Nature of the Scientific Community.* Yale, July 1970, unpublished.

————. *The State of the Art in Science Policy Studies.* Copenhagen, March 1972, unpublished.

Pye, L. W. *et al. Aspects of Political Change,* Little Brown, Boston 1966.

————. and Verba, S. *Political Culture and Political Development.* Princeton University Press, 1970.

Qadri, A. H. *Islamic Jurisprudence in the Modern World.* Ashraf, Lahore, 1973.

Qubain, Fahim I. *Education and Science in the Arab World.* Johns Hopkins Press, Baltimore, 1966.

Qurashi, A. I. *Islam and the Theory of Interest.* Ashraf, 1967.

Qurashi, I. H. *Education in Pakistan.* Ma'aref, Karachi, 1975.

Qutb, M. *Islam. The Misunderstood Religion.* Darul Bayan, Kuwait, 1967.

Rahman, A. *Anatomy of Science.* National Publishing House, Delhi, 1972.

Rahman, A. *et al. Science and Technology in India.* Indian Council for Cultural Relations, 1973.

Rahman, H. 'Law, Development and Fundamental Rights.' *Impact International Fortnightly* 5, 7, 8-9 (1975).

Rahnema, M. 'Iran: Science Policy for Development.' *Impact of Science on Society* 19, 53-61 (1969).

Ramadan, S. *Islamic Law: Its Scope and Equity.* Islamic Centre, Geneva, 1970.

Rao, S. R. 'An Example for the Third World.' *New Scientist* 59, 451-2 (23 August 1973).

Rashid, S. M. 'Health Problems.' *Hamdard* 17, 7-12 (July-Dec. 1976).

Ravetz, J. R. 'Criticisms of Science.' In D. de Solla Price and I. Spiegel-Rosing (eds.), *Science, Technology and Society.* Sage Publications, London and Beverly Hills, 1977.

————. *Scientific Knowledge and its Social Problems.* OUP, 1971.

————. 'What Can We Learn From The Freaks?' Paper presented at the CNAM Colloque, Paris, 4-6 December 1975.

Reader, D. H. *The Black Man's Portion.* OUP, Cape Town, 1961.

Reddy, A. K. N. 'Alternative Technology: A Viewpoint from India.' *Social Studies of Science* 5, 331-42 (1975).

Reich, C. *The Greening of America.* Random House, 1970.

Restivo, S. P. and Venderpool, C. K. (eds.). *Comparative Studies in Science and Society.* Merrill, 1974.

Rhodes, R. I. *Imperialism and Underdevelopment. A Reader.* Monthly Review Press, New York, 1970.

Riazuddin. 'Higher Education as an Essential Part of the Scientific Effort of a Developing Country and International Cooperation.' Address at the British Association Meeting, Durham, Summer 1970.

Roderick, H. 'The Future Natural Sciences Programme of UNESCO.' *Nature* 195, 215 (1962).

Rose, H. and Rose, S. 'The Incorporation of Science.' In Rose, H. and Rose, S. (eds.), *The Political Economy of Science: Ideology of/in the Natural Sciences.* Macmillan, 1976.

————. 'The Myth of Neutrality of Science.' In W. Fuller (ed.). *The Social Impact of Modern Biology.* R. & K. Paul, London, 1971.

Rosenberg, N. *Capital Formation in Underdeveloped Countries.* AER, 1960.

Rosser-Owen, D. G. *Social Change in Islam: The Progressive Dimension.* Open Press for the Muslim Institute, Slough, 1976.

Rostow, W. W. *The Stages of Economic Growth: a Non-Communist Manifesto*. CUP, 1963.

Rozak, T. *The Making of a Counter Culture*. Doubleday, New York, 1969.

————. 'Science, A Technocratic Trap.' *Atlantic* 230 (1), 56 (July 1972).

————. 'Science, Knowledge and Gnosis.' *Daedalus*, Summer 1974.

————. *Where the Wastelands End*. Doubleday, New York, 1972.

Roubiezck, P. *Ethical Values in the Age of Science*. CUP, 1969.

Russell, B. *Mysticism and Logic*. Allen & Unwin, London, 1910.

Rzoska, J. (ed.). *The Nile: Biology of An Ancient River*. W. Junk, The Hague, 1976.

Sagasti, F. R. 'A Framework for the Formulation and Implementation of Technology Policies. A Case Study of ITINTEC in Peru.' International Forum on Technological Development 1975, (unpublished).

Sahling, A. *Stone Age Economics*. New York, 1970.

Said, A. A. and Simmons, L. (eds.). *The New Sovereigns: Multinational Corporations as World Powers*. Prentice-Hall, New York, 1974.

Said, Hakim, M. 'Al-Tibb Al-Islami.' *Hamdard* 19 (1-6) (Jan.-June 1976).

Salam, A. 'Diseases of the Rich and Diseases of the Poor.' *Bulletin of the Atomic Scientists* 19, 3 (1963).

————. 'The International Centre for Theoretical Physics.' *Physics Today* 18 (3), 52 (1965).

————. 'The Isolation of the Scientist in Developing Countries.' *Minerva* 4, 461 (1966). Reprinted in E. Shils (ed.). *Criteria for Scientific Development, Public Policy and National Goals*. MIT Press. Cambridge, Mass. 1968, p. 200.

————. 'The Less-Developed World: How Can We Be Optimistic?' *New Scientist* 21, 139 (1964).

————. 'Memorandum on a World University.' *Bulletin of the Atomic Scientists*, March 1970.

————. 'A New Center for Physics.' *Bulletin of the Atomic Scientists*, December 1965.

————. 'Pakistan: The Case for Technological Development.' *Bulletin of the Atomic Scientists*, March 1964.

————. 'Science and Technology in the Emerging Nations.' In David Arm (ed.). *Science in the Sixties*. University of New Mexico Press, Albuquerque, 1965, p. 32.

————. 'Towards a Scientific Research and Development Policy for

Pakistan.' National Science Council of Pakistan, Karachi, 1970.

—————. 'The United Nations and the International World of Physics.' *Bulletin of the Atomic Scientists,* February 1968.

El-Said, M. E. M. 'A Study of National Science Policy Making Bodies of 61 Countries and a Brief Account of the Historical Development of The Science Council of UAR.' Unpublished n.d. (Copies available from the author at 101 Kasr El-Eini Street, Cairo, UAR).

Sardar, Z. 'Al-Biruni, 937-1048: Encyclopaedist, Scientist, Philosopher.' *Quest: Journal of the City University,* no. 27, Summer 1974.

—————. 'The Quest for a new Science.' Paper presented at CNAM Colloque, Paris, 4-6 December 1975; also Muslim Institute Papers - 1, Open Press, Slough, 1976.

—————. 'Quietly fights the Don.' *Impact International Fortnightly* 3 (2), 8-9 (1973).

—————. 'What Alternative to Arab Oil?' *Impact International Fortnightly* 3 (13), 7 (1973).

Sardar, Z. and Rosser-Owen, D. G. 'Science Policy and Developing Countries.' In I. Spiegal-Rosing and D. de Solla Price. *'Science, Technology and Society.'* Sage Publications, London & Beverley Hills, 1977.

Sartre, J. P. *Preface to Fanon: The Condemned of the Earth.* London, 1967.

Sayeed, K. bin. *Pakistan: The Formative Phase, 1857-1948.* OUP, London, 1968.

Scafeti, A. C. *Implications of Agricultural and Industrial Development.* Systems Development Corporation, Santa Monica, California, 1969.

Shils, E. 'End of Ideology?' *Encounter* 52 (8) (November 1975).

Scalapino, R. *Asia and the Road Ahead.* University of California Press, 1976.

Siddiqui, K. *Conflict, Crisis and War in Pakistan.* Macmillan, London, 1974. ———

—————. *Functions of International Conflict: A Socio-economic Study of Pakistan.* Royal Books, Karachi, 1975.

—————. *Towards a New Destiny.* Open Press, for the Muslim Institute, Slough, 1974.

Schneider, K. R. 'Development Universities, Special Institutions for the New Nations.' *International Development Review* 10, 17-22 (1968).

Schumacher, E. F. 'Economics Should Begin With People Not With God.' *The Futurist* 8 (6) (December 1974).

—————. *Small is Beautiful.* Abacus, London, 1975.

Schon, D. *Technology and Change: The New Heraclitos.* London, 1967.

Schone, K.J. 'Management Aid in Developing Countries.' *Quest: Journal of the City University* 29 (Spring 1975).

Schultz, T. W. 'What Ails World Agriculture?' *Bulletin of the Atomic Scientists*, January 1968.

Siddiqi, M. Shafqat Husain. 'Statement from Pakistan on Organisation and Management of Research and the Difficulties Experienced.' National Research Council, Karachi, n.d.

Siddiqui, M. N. 'A Survey of Contemporary Literature on Islamic Economics.' Paper presented at the International Conference on Islamic Economics, Mecca, 5-11 April 1975.

Siddiqui, M. Z. 'Islamic Culture—What Do We Mean By It?' In *Islamic Culture: A Few Angles*. Ummah, Karachi (undated).

Siddiqui, S. 'Problems Relating to the Utilisation of Research Results.' In W. Moorehouse (ed.). *Science and the Human Condition in India and Pakistan*. Rockefeller University Press, New York, 1968, pp.113-16.

Silcock, T. H. *Southeast Asian Universities. A Comparative Account of some Development Problems*. Duhe University Press, Durham NC, 1964.

Sinha, R. *Food and Poverty*. Croom Helm, London, 1976.

Sklair, L. *Organised Knowledge*. Paladin, London, 1973.

Skolisnowski, H. 'Knowledge and Values.' *Ecology* 5, 18 (January 1975).

Snow, C. P. 'The Two Cultures.' In *Public Affairs*. Scribner, New York, 1971.

Solomon, J. J. *Science and Politics*. Macmillan, London, 1973.

Sprey, J. 'The Problem of Choice.' In *Problems of Science Policy*. OECD Seminar, Jory-en-Josas, France, 19-25 February 1967.

Srientvason, K. 'Consultation as a Means of Technology Transfer.' International Seminar on Technology Transfer, Council for Scientific and Industrial Research, New Delhi, 1972.

Stepan, N. 'Science in a Developing Country.' *Beginnings of Brazilian Science: Oswaldo Cruz, Medical Research and Policy 1870-1920*. Science History Publications, to be published.

Stephens, R. *The Arabs' New Frontier*. Temple Smith, London, 1973.

Sulayman, A. H. A. Ali, 'The Theory of Economics in Islam: The Economics of Tawheed and Brotherhood, Philosophy, Concept and Suggestions for Policies in a Modern Context.' In *Contemporary Aspects.of Economic and Social Thinking in Islam*. The Muslim Students' Association of the US and Canada, 1970.

Sunkel, O. and Paz, P. *El Subdesarrollo Latinoamericano y la Teoria del Desarrollo*. Siglo XXI editores, Mexico, 1970.

Sussex Group. 'Draft Introductory Statement for the World Plan of Action for the Application of Science and Technology to Development.'

Annexe II to UN, *Science and Technology for Development: Proposals for the Second UN Development Decade.* New York, 1970.

Sutton, A. C. *Western Technology and The Soviet Economic Development 1930-1945.* Hoover Institution, New York, 1972.

Taylor, G. R. *The Doomsday Book.* Thames and Hudson, London, 1971.

Thorp, W. L. *The Reality of Foreign Aid.* Praeger, New York, 1971.

Tibawi, A. L. *English Speaking Orientalists.* Luzac and Islamic Cultural Centre, London, 1964.

————. *Islamic Education.* Luzac, London, 1972.

Tunaya, T. Z. *Islamcilik Cereyani.* Baha Matbaasi, Istanbul, 1962.

Turkcon, E. 'The Limits of Science Policy in a Developing Country: The Turkish Case — A Study Based on the Experience of the Scientific and Technical Research Council of Turkey.' *Research Policy* 2, 336-63 (1974).

Turkeli, A. 'Doctoral Training Environments and Post-Doctoral Productivity of Turkish Physicists.' UNESCO Document SC. 72/CONF. 3/3 Annex 24 (b), Ankara, 1972.

Turkey, Government of. *Second Five Year Development Plan 1968-72.* Central Bank of the Republic of Turkey, Ankara, 1969.

Ummah. *Studies in the Commonwealth of Muslim Countries*, Karachi (undated).

————. *The World Muslim Gazetteer.* Karachi, 1976.

United Nations, Advisory Committee on the Application of Science and Technology to Development for the Second United Nations Development Decade. *The Application of Computer Technology for Development.* UN–E34800, ST/ECA/136,1971.

————. *Multinational Corporations in World Development.* UN ST/ECA/ 190, 1973.

————. *Report on the First Session of the Committee on Science and Technology for Development.* UN–E/5272, 1973.

————. *World Plan of Action for the Application of Science and Technology to Development.* E/4962/Rev. 1 ST/EVA/146, 1971.

————. Department of Social Affairs. *Report of the World Social Situation.* New York, 1963.

————. *Science and Technology for Development. Report of the UN Conference on the Application of Science and Technology for the Benefits of the Less Developed Areas.* 8 vois. New York, 1963.

————. *Summaries of the Industrial Development Plans of Thirty*

Countries, New York, 1970.

UNESCO. *Moving Towards Change.* Paris, 1976.

————. *Natural Science Policies in Countries of South and South-East Asia.* Science Policy Studies and Document No. 3, Paris, 1965.

————. Programs for National Sciences and their application to Development. UNESCO SP/801/26/1086, Paris, 1973.

————. *The Promotion of Scientific Activity in Africa.* Science Policy Studies and Documents No. 11, Paris, 1969.

————. *Report of the Ad Hoc Working Group of the United Nations Advisory Committee on the Application of Science and Technology to Development (UNACAST) on UNESCO's Science Policy Programmes.* UNESCO/93/EX/14, Paris, 1970.

————. *The Role of Science and Technology in Economic Development.* Science Policy Studies and Documents No. 18, Paris, 1970.

————. *Science and Technology in Asian Development.* Paris, 1970, pp. 21-3.

————. *Scientific Research in Africa, National Policies, Research Institutions.* Paris, 1976.

————. *World Summary of Statistics on Science and Technology.* Statistical Reports and Studies No. 17, Paris, 1970.

Usmani, I. H. 'Developing Countries and IAEA'. Paper delivered at 13th Session of the General Conference of IAEA, Vienna, 1969.

————. *International Atom in the Seventies.* Speech at the 14th General Conference of IAEA. Associated Printers and Publishers, Karachi, Pakistan, 1970.

————. *Organization and Financing of Scientific Research in Pakistan.* Pakistan Atomic Energy Commission, Karachi, 1971.

————. *Search for Research.* CENTO Symposium, November 1964.

Vaitsos, C. *Patents Revisited. Their Function in Developing Countries.* J. Dev. Stud. 9, 71-97 (1972).

Victorizc, T. 'Diversification, Linkage and Integration. Focus in the Technology Policies of Developing Countries.' Seminar on the role of small-scale industries in the transfer of technology. OECD, June, 1973.

Ward, B. *et al.* (eds.). *The Widening Gap.* Columbia University Press, 1971.

Ward, P. 'Indonesian Libraries Today.' UNESCO Bell Libr. 29 (9), 182-7 (1975).

Waterston, A. *Development Planning: Lessons of Experience.* OUP,

1966.

Wellisz, S. 'Lessons of Twenty Years of Planning in Developing Countries.' *Economic Quarterly* 38, 128 (May 1971).

Wells, L. T. 'Economic Man and Engineering Man. Choice of Technology in Low Wage Country.' *Economic Development Report,* Autumn 1972.

Westfall, S. 'Newton and the Fudge Factor.' *Science* 179, 751-8 (1973).

Whitaker, C. S. *The Politics of Tradition,* Princeton University Press, 1970.

White, L. 'Historical Roots of our Ecological Crisis.' *Science* 155, 1203 (1967).

White, L. A. 'The Concept of Culture.' In Ashley, M. F. (ed.). *Culture in the Evolution of Man.* OUP, 1962.

Wilkinson, J. *Technology and Human Values.* Centre for the Study of Democratic Institutions, Santa Barbara.

Wilkinson, P. *Terrorism versus Liberal Democracy. The Problem of Response.* Institute for the Study of Conflict, London, 1976.

Wilkinson, R. G. *Poverty and Progress.* Methuen, London, 1973.

Wilson, C. W. *Technological Development and Economic Growth.* Indiana University Press, 1971.

Wineberg, S. 'Reflections of a Working Scientist.' *Daedalus.* Summer 1974.

El-Yacoub, H. 'Sudan's Five-Year Plan for Research/Development.' *Impact of Science on Society* 26 (2/3) (1976).

Zaheer, S. Husain. 'The Development of Science and Technology in Underdeveloped Countries.' *Scientific World* 12, 9 (1968).

Zahlon, A. B. 'Science and Higher Education in the Arab World.' *New Edinburgh Review* No. 23, 12 (1973).

————. 'The Science and Technology Gap in Arab/Israeli Conflict.' *J. Palestine Studies* 1, 17-36 (Spring 1972).

————. 'Science in the Arab Middle East.' 1967, unpublished.

————. 'Migrations of Scientists and the Development of Scientific Communications in the Arab World.' 1969, unpublished.

de Zayor, P. G. *The Law and Philosophy of Zakah.* Al-Jeddah Press, Damascus, 1960.

Zijderveld, A. D. *The Abstract Society.* Allen Lane, London, 1972.

Ziman, J. 'Some Problems of the Growth and Spread of Science into Developing Countries.' *Proceedings of the Royal Society* A311, 349-69 (1969).

————. *Public Knowledge.* CUP, 1968.
————. *The Source of Knowledge.* CUP, 1976.

INDEX

For Product Safety Concerns and Information please contact our EU
representative GPSR@taylorandfrancis.com
Taylor & Francis Verlag GmbH, Kaufingerstraße 24, 80331 München, Germany